D1016900

WHAT'S *Eating* YOU?

WHAT'S *Eating* YOU?

PEOPLE AND PARASITES

EUGENE H. KAPLAN

*Illustrated by Susan L. Kaplan
and Sandy Chichester Rivkin*

PRINCETON UNIVERSITY PRESS
PRINCETON AND OXFORD

Library of Congress Cataloging-in-Publication Data

Kaplan, Eugene H. (Eugene Herbert), 1932–
   What's eating you? : people and parasites / Eugene H. Kaplan ; illustrated by
Susan L. Kaplan and Sandy Chichester Rivkin.
       p. cm.
   Includes bibliographical references and index.
   ISBN 978-0-691-14140-4 (hardback : alk. paper) 1. Parasites—Popular works.
2. Host-parasite relationships—Popular works. I. Title.
   QL757.K365 2010
   616.9'6—dc22   2009021742

British Library Cataloging-in-Publication Data is available

This book has been composed in ITC New Baskerville STD
Printed on acid-free paper. ∞
press.princeton.edu

Printed in the United States of America

10 9 8 7 6 5 4 3 2 1

The World Toilet Summit, which just took place in New Delhi, is . . . pushing for better sanitation technologies.

. . . Diarrhea kills 1.6 million children each year—more, even than malaria—and the pollution of drinking water with waste is the principal problem.

Experts all agree that the two most important public health measures in the world, measures that saved more lives than either vaccines or antibiotics, have been in place since the time of the Roman Empire. . . . [They are] running water and toilets. . . .

. . . lack of adequate toilets threatens more children than, for example, global warming does.

No movie star has died of diarrhea. No politician has died of poverty.

—DONALD MCNEIL, JR.
(MODIFIED FROM A QUOTE BY JACK SIM), 2007

CONTENTS

PREFACE

PERSONAL PARASITES

LET'S GET THIS STRAIGHT. I do not feel that I should have deliberately infected myself to enhance my teaching skills. My infections were accidental. The era of self-sacrifice is over. It is told that the famous German pathologist Theodor Bilharz placed some cercariae (the infective form of schistosomiasis) on his stomach and took notes as they burrowed through his skin, eventually to lodge in his liver and lower mesenteric veins. This parasite is found in Egypt and the disease it causes is named after him, bilharziasis (bilharzia). I had no such heroic intentions when I suffered through amebic dysentery in Egypt or when I "gave birth" to a worm the length and thickness of a pencil, which I picked up in Israel, but their stories certainly enhance and enliven my lectures.

This is a book of stories about parasites. Some stories describe my own experiences, some have been told to me. A thread runs through them: entwined in the twists and turns of DNA are a dual set of the miraculous mechanisms of evolution. Why a *dual* set? The evolutionary steps of the host are matched by those of its parasites in a grim dance of death. Should one species make a misstep, it becomes extinct. Should the parasite become so harmful that it kills its host, the host species disappears; should the host evolve a perfect defense against the parasite, the parasite becomes extinct. In effect, there is an uncomfortable evolutionary accommodation between the partners in this intimate interrelationship.

Evolution can be defined as a series of processes that lead to more and more successful adaptations of a species to its environment. To elucidate these adaptations it is often necessary for the parasitologist to wallow in feces.

As in all evolutionary/ecological endeavors, one must examine habitats to clarify trophic relationships. Unfortunately, these habitats are the guts of host animals—and their effluvia. Stories herein describe midnight forays into intestines, large and small, to find parasites hiding therein. The almost overwhelming stench that accompanies such an endeavor is matched only by the unpleasant examination, under the microscope, of the feces. In this book I will describe this reality so that you can have the most authentic experience.

WORMLY

In order to make real the world of parasites, I require that my students examine human feces to identify the eggs of the parasites that afflict most of humankind. This is a tough order for both students and professor. My problem is that I cannot get enough infected feces to fulfill my requirements. Fortunately, I had a dog named Wormly by the veterinarian when she was first brought from the pound for the usual puppy inoculations. I was ecstatic at her wormy opulence, but she was dewormed—my family refused to accommodate my professional needs. She made her contribution in a different way. It is possible, in this best of all worlds, to send away for vials of eggs of human parasites. But the sterile professional vials contain eggs filtered from feces and preserved in alcohol. How to create the proper fecal experience for the students? Our dog was required to make the ultimate contribution: she provided the feces. In the name of educational integrity, I mixed the human parasite eggs into the dog's feces under the assumption that no one would be so sophisticated as to be able to differentiate between human and dog feces. The vials were numbered and the students would get one point on the midterm for every egg they properly identified. Students will do anything for points on the midterm, and they did the necessary wallowing in the feces.

Good experience.

This book is a compendium of lurid stories told to generations of young people to keep them awake so that they would listen to what followed—lectures filled with minutia, the manifest burden of science students.

ACKNOWLEDGMENTS

THIS BOOK IS DEDICATED TO MY LOVING FAMILY. Not content with providing inspiration, daughter Sue created drawings for the book. Daughter Julie edited the legends. Wise wife participated in many of the adventures described herein.

My loyal and talented laboratory assistants made it possible to challenge students with their advice, example, and tedious, often nauseating, labor in the preparation of living materials for the course. My deepest gratitude goes to Debra Bradley, perennial assistant in field and lab, and numerous others over the years.

Bob and Arn Johnson, "hired guns" and mighty hunters, helped obtain specimens.

My thanks to Professor Paul Billeter for reading sections of the manuscript. Alan Shuman provided funding for my aquaculture research. David Wallach, partner in the fish farm in Côte d'Ivoire, and Meir Feuchtwanger, founder of Project MATAL (Israel), provided wise counsel.

Above all, I take this opportunity to express my gratitude to the hundreds of students who took the parasitology course for their good humor, intelligence, and tolerance. My life among these people was a delight.

The warm, kind, and competent production and editorial staff at Princeton University Press, Mark Bellis, Dimitri Karetnikov, and Maria Lindenfeldar, literally made this book. They are wonderful.

Special thanks to Robert Kirk, senior editor for biology and natural science, for seeing merit in this book and its predecessor.

This book has been "purified by fire." Lyman Lyons, with sensibility, tact, and fine editorial skills, has seen to that.

Don Duszynski has been a source of support and a steadying hand when this project was in doubt. I am very grateful for his dedication to parasitology, manifest, in this instance, by his willingness to edit the text for accuracy.

I used the extraordinary textbook *Foundations of Parasitology*, coauthored by Larry Roberts and John Janovy, Jr., as a source of human-interest material. The authors' ability to intersperse fascinating stories with highly technical content was inspiring.

When the chips were down and help was needed desperately, Lillian Mayberry and Jack Bristol came to the rescue with their voluminous expertise in parasitology.

The production of every book has its emergencies. Sue Kaplan and especially Sandy Chichester Rivkin worked under great stress to produce the beautiful plates that grace this book.

Most of the events in this book are real except when written in the third person. The characters are not. They are composites of people I have known, mixed in with a ghostly population of characters from parasitology lore. Any resemblance to people, living or dead, is coincidental and not intentional.

The mistakes in this book are solely my responsibility.

APOLOGIA[1]

ANTHROPOMORPHISM and TELEOLOGY, two of the mad horsemen of the biological apocalypse, run rampant through the pages of this book. Anthropomorphism is to give human characteristics to animals or objects. ("The Little Engine That Could" is an example of starting children off on the wrong logical foot.) For example, abundant references have been made to stiletto-nosed vampires (mosquitoes) and fiendish tapeworm monsters. The patently prurient nature of the stories herein seemed to require committing this sin to make them more interesting.

Teleology is confusing cause with effect. Any implication that an animal has evolved to acquire a biological niche (e.g., tapeworms evolved their complex life cycles in order to find a means of entry into the human body) is an example of poetic license. I have railed against these facets of illogical thinking in another book,[2] and I apologize for resorting to them in the name of storytelling.

ON THE SACREDNESS OF LIFE

God blessed them, saying unto them, "Be fruitful, multiply, fill the
earth and conquer it.
Be masters of the fish of the sea, the birds of heaven and all living
animals on the earth."

—GENESIS 1:28

I AM AN ANIMAL LOVER. Any naturalist is—but I believe I have a balanced approach to living things. I revere life, but in order to further learning and research, I am willing to sacrifice animals. I search for suitable animals. I use animals that are considered "vermin." But they are living, nevertheless. Is a cockroach any less worthy, less "living," than a human; is a worm? Is there justification for sacrificing the life of roach or worm toward the admittedly remote possibility that such a sacrifice will serve humanity?

On our campus exists an organization that has as its raison d'être to save animals that are used for research, on the grounds that all life is sacred and cannot be sacrificed for any purpose. A representative of this vigorous group threatens to close down the university animal facility if researchers kill fruit flies for genetic research.

There is another group that feels that it is not acceptable to sacrifice an aggregation of even a few cells that do not possess the slightest physical or functional suggestion of a nervous system. Even a single fertilized egg is sacred and should not be used to further research that promises to help cure human ailments. Thus, the one-celled human precursor, the fertilized egg, should not be killed to save from death thousands of fully functioning humans.

This book is not for them.

It is not for me to defy such a philosophy. If you are of such disposition, do not read on.

I believe that students must not learn biology in a vacuum. *Life in its complexity cannot be represented by pickled organisms in bottles* if progress is to be made. It is unnatural to make believe one is dealing with life if death is a model. Future researchers and healers need to get a feeling for the sacredness of life by appreciating the miracle of the living body. Consequently, only living organisms are ultimately suitable for advanced study. My course in parasitology uses live (freshly sacrificed) cockroaches, mice, frogs, and seagulls to give the student research experience. It is illogical to prepare a student of biology for a career in investigating or curing living things without touching living things. It is true that a molecular biologist can spend a lifetime being handed cells or tissues by a technician and never see a live animal—but the cells being used for his or her research must have originated from a sacrificed living animal, so the subject is moot.

Computers can simulate and model tiny aspects of life, but they are as valuable for teaching and research as pickled frogs. They lack the unpredictable—the infinite variability of living systems.

When I contemplate forty-plus years of teaching students about parasites, I think of all the physicians and researchers that saw their first beating heart when they searched the thoracic cavity and lungs of a frog for flatworm parasites. The ultimate moment of truth is when a student investigates a tiny, clotlike mass in the posterior mesenteric veins of a seagull and teases the tissue away to find a living schistosome, the uncured curse of Africa.*

Most American medical schools do not require courses in parasitology.† If they do, they represent the field as a static sequence of slides and lectures. What about the dynamics of the affliction, its complexity, the great variability of symptoms?

Fortunately, it is in the nature of parasites to keep the host alive. The symptoms accompanying Aunt Bessie's arrival home from a trip to Mexico will eventually respond to the confused groping of her family doctor, who tries one potentially lethal medication after another in the quest for a cure.

How many of my parasitology students chose a career in research after helplessly watching the red blood cell count of a mouse decline

---

* Cures exist, but they are not accessible to most infected people.
† "According to a survey of 41 medical schools in the U.S., 51.3% do not require a course in parasitology and 35.6% offer no parasitology courses."[3]

to eventual death from malaria?* I hope that one of these hundreds of students will discover the cure for a parasitic disease or will have the wherewithal to cure a patient afflicted with a parasite.

* The malaria we investigate is mouse malaria, *Plasmodium berghei.* It is not transmissible to humans. The species of schistosome we study is specific to birds. It cannot cause a response in humans other than a rash—and then only under circumstances different from those in the classroom. We use no more than eight seagulls and sixteen mice each year. The mice are bred for research purposes, and the seagulls are collected under license from federal, state, and local authorities.

INTRODUCTION

THE SALINE SOLUTION—
AN INNER SEA

THERE IS ANOTHER SEA, a dark red ocean of blood filled with monsters no less threatening than a shark or venomous blue-ringed octopus. For the majority of humans on the Earth, this inner sea is populated with dangerous beasts—parasites. They suck the blood from the outside or use it as a habitat, drawing sustenance from the human body and wringing from it life-sustaining forces that they pervert to their own needs.

Parasites can flow with the blood to take haven in the richly oxygenated lungs, suckling on the inside of the breast rather than the outside, to prepare themselves for a lifetime of harm to the host. Some have adapted to take advantage of this pulmonary paradise to go through their developmental stages, a hiatus permitting ever more adaptive efficiency on the road toward the creation of the *perfect parasite*. Perfection is not necessarily physical—it can be physiological or biochemical. The tapeworm, lacking a mouth, an intestine, or an anus of its own, appears to be an unprepossessing piece of segmented linguini. But its outer covering is feathered into microscopic microvilli, making the surface fuzzy like a blotter, able to differentially soak up molecules of nutrients. It lives in the small intestine. There it is bathed in a river of fluid composed of chewed up chunks of lunch mixed with digestive juices to make a rich soup containing enzymes that chemically shred molecules of food into their constituent building blocks. The Whopper you ate is reduced into amino acids by your proteolytic enzymes. *But the tapeworm is made of proteins*—how is it not digested with the food? Not only does the multifunctional surface facilitate the passage of nutrients into its body, but it also prevents proteolytic enzymes from destroying it by producing an antienzyme.

Many parasitic nematodes (roundworms) spend time in the lungs, growing and changing, like the pupae of butterflies. Once mature, they leave their nursery and drift in the blood to their final destination: the intestines, the lungs, lymph spaces of the groin, or under the skin. The story is told of a British expatriate who, at the end of a long tenure in Sudan, returned home to the traditional retirement destination, the hallowed ivy-covered cottage in Surrey. Every day, precisely at teatime, a tiny worm would swim across his cornea, visible to his guests. He became quite the popular host. The worm was *Loa loa*, a sometimes benign relative of the filarial worm that causes the grotesquely deformed legs and testicles characteristic of elephantiasis.

But why did this worm appear precisely at 4 p.m. every day? Why was the worm on a timetable like a railroad? Forty years of conjecture have yielded this explanation: The man's blood was thinned by years of living in the blinding heat of Sudan. Consequently, he must have lit the fireplace at teatime to take the chill out. The warmth of the fire absorbed by his skin must have brought subcutaneous worms close to the surface. They undulated under the skin, through the sclera and across the cornea of the eye. That's the best I can do to explain the mystery!

RELATIONSHIPS

Relationships are invariably complex. An intimate long-term attachment can suddenly go awry; one partner can turn on the other, with dire consequences. Often, over time, the dependent becomes the dominant. Sometimes the relationship becomes so all-encompassing that one member is almost subsumed by the other. Separation is rampant as youthful indiscretions cause breakups. Often, to everyone's surprise, one member discovers he/she has a penchant for members of the same sex. Often the discovery is made when a partner becomes a flamboyant cross-dresser.

Am I talking about people? No. The following stories are gossip not about humans, but about animals that live together "for better or for worse." Some have long-term interdependencies, and their mutual survival depends on the interaction. Some relationships are ephemeral, and one or another partner can wander off at any time, both being capable of independent living. Sometimes one member gets something from the other but does not apparently contribute anything to the partner's well-being—the stay-at-home parent, for example.

Sometimes one partner hurts the other.

Relationships can change. If one animal receives shelter from another, the interaction might be commensalism (one partner benefits while not harming the other). But careful observation of the partners over time can reveal that the apparent beneficiary betrays its protective host by biting chunks of tissue from its living haven. In a Jekyll-Hyde switch, the commensal changes into a parasite.

Hypothetically, there is a sequence of developmental stages leading to the dazzling complexity of the host-parasite relationship. First, an accidental association. Then a progression to the ultimate intimacy, one species taking sustenance from the other, to the host's detriment. But parasitism is not the only interactive evolutionary process—often there is a succession of less inclusive interactions between animals of different species. Symbiosis (living together) has three major categories—mutualism, commensalism, and parasitism. To oversimplify:*

Mutualism (formerly called symbiosis) is an interaction between two organisms, each of a different species, both of whom benefit.

Commensals derive something out of the relationship but do not harm the host.

Parasites harm the host.

MUTUALISM

Mutualistic relationships are likely to be ancient and long term since each participant has had to evolve mechanisms not only for its own benefit, but for its partner's. Corals, living in seas virtually devoid of food (plankton), cannot survive without hosting plantlike one-celled

---

*For more accurate definitions, see the glossary.

organisms, zooxanthellae, in their tissues. The zooxanthellae are able to perform photosynthesis to produce food, shared with the coral partner. The coral, in turn, contributes its waste, carbon dioxide, to the zooxanthellae, which they use to manufacture extra food, making it possible for *them* to survive.

The zooxanthellae benefit further—the corals protect them. The corals get another benefit as well: the extra energy provided by their zooxanthellae makes it possible for the corals to extract calcium carbonate from the water to manufacture their famed chalklike skeletons. How's that for complex interdependence?

Similarly, the cleaning goby, *Labroides dimitiatus*, gets exclusive permission to graze in a flourishing field of parasites on the body of its fishy partner, who benefits by being relieved of its tissue-eating and bloodsucking burden.

COMMENSALISM

One partner benefits from the relationship and the other neither benefits nor is harmed. Commensalism means "to eat at the same table." The remora clings to the belly of the shark and feeds on the scraps of its meal. An ameba, *Entameba gingivalis*, lives in the human mouth. It inhabits the gum line, harmlessly eating bacteria between the teeth. It gets sustenance from its partner but has not become a parasite. Would it be beneficial for it to become a parasite? Why should it? The ameba lives happily in the mouth, protected by its human partner, not causing the slightest damage to its human haven. Is this not better than parasitism, where even minimal harm may ultimately impair the host, threatening the existence of the parasite?

PARASITISM

Parasite and host often evolve physiological accommodations to one another. The parasite acts to keep the host as healthy as possible while surreptitiously extracting the wherewithal to sustain its life. This is logical and intuitive. But more incomprehensible are the *behaviors* that lead to interactions between parasite and host. How did they evolve?

How do bird brood parasites figure out what nests to lay their eggs in? How do parasites develop behaviors parallel to their host's to allow them to insert their developmental stages at a particular time? The kentrogon larva of the barnacle, *Sacculina*, will die after a week of aimlessly searching if it cannot find the leg joint of a newly molted crab into which it injects is essence, stem cells that will develop into a cancerlike invader whose tendrils insidiously permeate the insides of the crab, keeping it alive until the last moment, like the monster in the movie *Alien*.

It is the evolution of parasite behavior that so fascinates.

All three relationships are found in abundance in nature. One might infer that less intimate interactions are precursors to more precise relationships between host and parasite. For example, a tiny white pinnotherid crab lives in the mantle cavity of a clam. It feeds on the mucous-entangled planktonic waste of the clam. But every once in a while, the crab takes a nip from the clam's mantle tissues. Is this part of a transition from commensalism to parasitism? Perhaps.

Every so often an exciting event occurs in the life of a parasitologist. A zebra in the local zoo died. The zookeepers had standing instructions to let us know of any deaths that occurred. While this was a sad event for them, it was a source of jubilation to our necrophiliac crowd. We arrived soon after the unfortunate event. Without further ado, someone whipped out a butcher knife and we began diving into the rumen (stomach chamber) of the newly dead zebra. It was filled with an almost infinite number of flagellated protozoans. Was this a sick animal, done in by intestinal parasites? No, the protozoans were mutualists; they helped digest the tough grass that the zebra ate and received a warm, moist, protective environment for their trouble. Yet closely related flagellated parasitic protozoans found in the human vagina and female reproductive system can cause irreparable harm.

There are many mysteries in the depths of the inner sea. We will discuss a few monsters of these bloody depths.

1

LAND OF SMILES

STANDING ON A STREET CORNER. Crowded, bustling. Watching a street vendor grilling what seems to be hot dogs. Hungry, I move closer—and recoil in shock. Grotesquely hanging from the edges of the bun are masses of blackened, rigid strings. Customers walk away munching with gusto, spitting out the strings as they walk. Making believe I am still hungry, I saunter up to the grill. Steam rises from charred corpses with elongate bodies and ten burnt legs. A pair of long, skinny appendages projects frontward, revealing this streetside delicacy to be the giant Malaysian prawn, *Macrobrachium rosenbergii.* It seems to be the Thai version of fast food.

This is Bangkok, not New York—or is it? Lined up at the stand are grungy American kids munching with some satisfaction on this exotic food, reveling in the obvious disgust of some of their brethren. The kids are part of an army of recent college graduates roaming the world, seeking the meaning of life in some other-worldly philosophy—seeking their guru in this land of saffron-robed monks. Some are pure vegetarians. These vegan vagrants are required by some inner directive to obtain their carbs as noodles from an adjacent noodle stand. The noodles are fascinating—ivory colored little balls, coils, and flat ribbons. They were plopped into boiling water for a minute, strained out steaming, and placed on a small plate. They look pale—and delicious.

Thais are, to me, distinguished by three characteristics: ubiquitous smiles, kindness, and cleanliness. The cleanliness is well intentioned but illusory. Take a water taxi up the Chao Phraya River that borders Bangkok. Houses jumble onto one another along the banks of canals called "klongs" and along the riverside. Large, amphora-like urns on each porch catch pure rain, for the river is dark with mud eroded from lumbered-over hillsides upriver, making it like thick soup. The vegetables are spoiled remnants from the markets; the meat, dead rats. Occasionally human waste punctuates the garbage-flecked surface.

Rainwater must be conserved for drinking and cooking. Children are given their daily bath with river water and scrubbed until shining—with the utterly polluted water.

The college kids on the corner go off to their dollar-a-night hostels and we go back to the Bangkok Palace to cool off and watch in-house Arnold Schwarzenegger videos—nothing else is available. In the evening we are ready for some night life. Across the street is a huge building. On its roof is a red neon sign indicating, in unintelligible Thai, its purpose. But alongside the garish sign is another, equally garish. It is a translation in neon: "FAST FOOD"—the Thai version of McDonald's. "Let's try it," I say. Wise wife points out that it is forbidden to eat regular Thai food except in the best restaurants. I insist. I go to the counter, behind which is a smiling woman. I point to a plate of particularly inviting flat noodles. I say, "What harm can noodles cause?" Wise wife says, "They will kill you."

The smiling woman picks up the plate of noodles. I reach for it. She turns around and places the plate into a grimy hand that miraculously extends from a black hole in the wall. "See, the noodles are going to be boiled," I say uncertainly. After a moment the woman hands me the newly heated plate. I cough up a few baht (cents) in payment and sit down at a table. Wise wife repeats, "They will kill you." I repeat, "The noodles are boiled." I swallow a noodle, noting that it is lukewarm.*

---

* Noodles cannot pass on amebic cysts, but human hands besmirched by fecal remnants can. The contaminated man behind the scenes was supposed to boil the noodles, but he heated them to a lower-than-boiling temperature and touched them, transferring the cysts. The warmth of the noodles served only to stimulate the encapsulated amebas.

Two days later, doubled over in pain, I alternate copious diarrhea with uncontrollable coughing. I think, "This cough is symptomatic of a bacterial infection," and we continue our trip. I almost fall off an elephant in the North. Later on, in Egypt, I am too weak to explore the magnificent temple at Karnak. We reach Israel. Wise wife insists that I bring a fecal sample to the "American Clinic" (distinguished from the others only by the fact that they charge American prices). I am at work. The physician calls. He says that I have "many, many, many amebas" in my intestines—amebic dysentery!

Those rotten American kids eat their way through Thailand and remain healthy. I eat one plate of noodles and get dysentery.

KILLERS IN THE GUT

The cause of amebic dysentery is the one-celled, amorphous protozoan called *Entamoeba histolytica* (histo = tissue, lytica = to dissolve). It really dissolves tissues, digesting its way through the walls of the intestines, in the process greatly irritating them. This stimulus produces a massive outflow of neuronal activity that the brain converts into return messages increasing the motility of the intestines. The contents of the gut move so fast that the colon cannot absorb enough water from the feces, resulting in diarrhea.

Death from diarrhea-induced dehydration is the primary cause of infant mortality in the world. Frequently the diarrhea is caused by *E. hystolytica*. When in diarrheic (liquid) feces, the amebas take their usual form—or their usual nonform. Amebas are shapeless, single-celled organisms. Bulges appear and disappear with no apparent pattern. When a food particle touches the cell membrane, it is surrounded by these bulges (pseudopodia). Presto! The food is inside the animal and being deluged with enzymes. Digestion occurs. During this feeding phase, called the trophozoite, each ameba is able to reproduce by splitting in half, becoming two—part of a reproductive strategy that will eventually create a veritable invading army. Due to its amorphous shape, *E. histolytica* is not easily diagnosed. One criterion for differentiating this pathogenic species from other (harmless) intestinal amebas is whether or not a dot in the nucleus is centrally located or off center. Diagnosis is therefore difficult, though crucial. One must go through a regimen of possibly toxic medications if the dot is central, but if it is askew, the ameba is harmless and no medication is needed. Quite a responsibility for the technician half asphyxiated from the smell of the material under the microscope.

Entamoeba histolytica enters the body in contaminated water or food containing cysts that are resistant to the acids of the gastric juice. Normally, this very acidic fluid, secreted by the walls of the stomach, acts as a fluid gate, preventing the entrance of harmful organisms by killing them with acid (which will eat away a penny in a day). But some protozoan parasites produce plasticlike "shells," microscopic spheres called cysts, that resemble minute eggs. The impervious cysts pass through the gastric gate, entering the small intestine. They flow through the richly inviting small intestine, waiting until they reach the environs of the large intestine (colon or bowel) to escape their imprisoning capsules—but not before reproducing again, splitting into four or eight amoebulas. This massive geometric progression produces the inner army that is destined to do the damage.

The immature amebas escape from their protective prison and begin to feed on the copious food surrounding them. What causes an ameba to forsake its fetid paradise and invade the walls of the large intestine is unknown, but ameba genetics and host food habits are involved. Some human populations are particularly vulnerable. An epidemic in Durban in the 1930s killed thousands of black South Africans while hardly harming the white population. The diet of the poor, rich in carbohydrates, has been shown to stimulate the invasive mode of feeding. In theory, adequate protein in a human diet will satisfy the needs of the parasite. Or is it that protein-poor diets lead to malnourishment and the concomitant vulnerability to penetration of the gut wall? Or is it the host's genetic vulnerability to the pathogen?

INVASION OF THE CLONES

When an invading colony is established, it digests its way through the inner surface layer (mucosa) of the intestinal wall, becoming an army by splitting into an overwhelming number of clones. These legions enter the blood-vessel-rich submucosa. Like Roman armies, some sail off in the blood to establish outlying colonies elsewhere, primarily in the liver, sometimes causing fatal hepatitis. The rest of the legion digests its way through the submucosa and reaches the tough, muscular middle layers of the intestine wall. They are slowed but not stopped. In cross section the lesion looks like a microscopic mushroom as the frustrated amebas spread out under the seemingly impregnable muscle layer. Finally a narrow tunnel is drilled through, allowing the invaders to penetrate through the outer layer of the gut and burst out into the peritoneal cavity in a foul mixture of intestinal bacteria, feces, and amebas—often causing fatal peritonitis.

But in most cases the amebas remain in the colon and are carried along by the slowly flowing peristaltic river of intestinal contents. They confront obstacles and eddies where the flow is reduced. There the amebas establish their colonies. The cecum, a blind sac near the entry point of the small intestine into the large (from which the appendix protrudes), is one such vulnerable sluggish spot.

One can map the high-risk regions based on the flow rate. The sigmoid flexure, the S-shaped convolution before the rectum, is another area of colonization. Once the amebas reach the end of the "river" and pass out of the anus they are still dangerous. They may invade from the outside, producing surface colonies that penetrate the skin around the anus, then float off in the blood. Any organ in the body can be infected by migrating amebas in the blood: the eyes, the brain, the lungs, the liver.

During the first half of the twentieth century, few medications were available to cure intestinal protozoan infections. Traditional choices were similar to cleaning fluid—on the theory that the poison would kill the tiny parasites before they killed the bigger host. Then a new drug was developed to kill flagellated parasitic protozoans infecting horses and cattle (there is a very profitable market for veterinary medicines). Incidentally, it was found effective against a variety of human parasites.

The horse medicine worked. In a few days my symptoms disappeared.

FROM OOZE TO INTESTINES

Once upon a time, amebas oozed their aimless paths through the mud of a pond. The seasons inexorably brought their changes. Spring became summer and the heat of the sun warmed the water. Warm water evaporates. The pond began to shrink and became a pool, then a puddle. Stressed by summer's changes, things died and rotted. Levels of ammonia associated with rot increased. Bacteria gorged themselves on the dead, reproduced massively, and used up most of the oxygen.

The pool continued to shrink; conditions became unbearable. Finally a smelly, muddy stain remained where the pond used to be. All its inhabitants died—or adapted to this most stressful of all environments. The most challenging stressor to overcome was dryness—to survive in the period of parched mud and wait out the drought until the rains came.

Over eons the amebas changed. Some evolved "shells"—microscopic invulnerable spheres called cysts that remained in the muddy bottom until autumn's reprieve. The rains returned and the pond revivified. Inside the cysts life stirred. But the amebas could not escape their self-imposed prisons. They had survived the dryness only to die in their newly evolved cysts in a futile evolutionary experiment. More almost limitless time passed. Finally, responding to signals in the water, a new kind of ameba cracked open its cyst and a new generation escaped into the all-providing mud.

What does this story tell? It seems to describe the evolution of pond-dwelling amebas. But there is a hidden meaning. Conditions in the drying pond very closely resemble the contents of the human intestine: high levels of ammonia and methane, low oxygen concentration; crowds of bacteria thriving at warm temperatures; the massive presence of organic matter.

But hell lies in the way to paradise: the burning cauldron of the pit of the stomach, its core so acidic as to kill all who pass that way. The amebas are thwarted. *But some species had evolved the safety of cysts.* They could pass through the acidic fires of hell unscathed. The same kinds of signals that saved them in the pond saved them here. They passed

PLATE 1

A. TRUTH, by Jules Joseph Lefebre (1870), here symbolizes the light of science finally shining on the many parasitic diseases afflicting the developing world. Poverty has accentuated the woman's problem (she cannot even afford clothes). She is suffering from advanced amebiasis seen in lesions in her cecum and colon.

B. CYST OF *Entamoeba histolytica.* Round; impervious wall. Cyst ingested in contaminated water, food, begins reproducing: divides into four nuclei, then excysts; cytoplasm pinches into four amebulae.

C. TROPHOZOITE of *Entamoeba histolytica* active feeding stage. Identified by dot (karyosome) precisely in center of nucleus. Spaces in cytoplasm are food vacuoles that often contain red blood cells. Feeds by extending pseudopodia around particles that end up inside food vacuoles.

D. CECUM OF LARGE INTESTINE OF WOMAN showing advanced lesion (black hole) containing thousands of amebas. Cecum opens into transverse colon (cut) at top, crossing above small intestine to descending colon and rectum. Flow slows at S-shaped curve at beginning of rectum allowing amebas to obtain foothold.

E. CROSS-SECTION OF WALL OF CECUM inside (lumen) at left. Colony penetrates lining (mucosa), expands into soft submucosa containing donut-shaped arteries; thin walled, oval veins that can be penetrated; amebas float in blood, can end up anywhere in body (often liver where they establish daughter colonies). Colony eventually penetrates through tough muscle and invades peritoneal cavity causing peritonitis. Amebas are dots at right bursting into peritoneal cavity. If the woman is not treated soon, she will die.

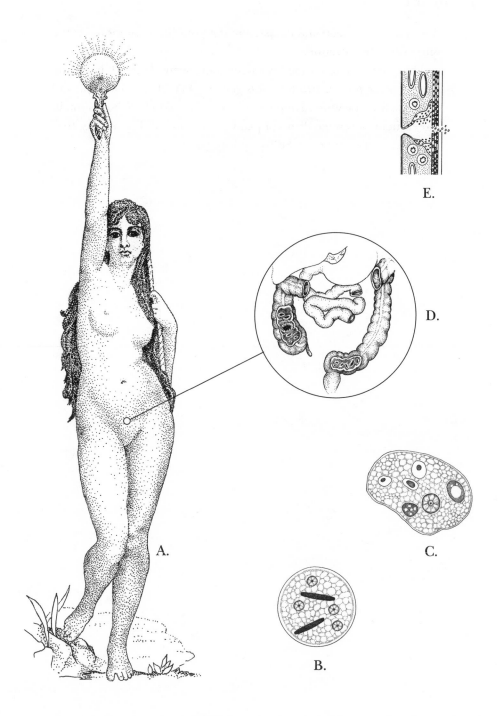

A.

B.

C.

D.

E.

through the stomach and excysted in the pondlike conditions on the other side, the intestine.

This phenomenon, called preadaptation, seems fanciful, but living models make it real. Two harmless genera of ameba, *Vahlkampfia* and *Sappinia,* live in sewage. They are able to encyst. Bad little boys swim in the effluent of sewage disposal plants. Laughing and splashing, they ingest the water and cysts. Still skeptical?

2

AN ENCOUNTER WITH
JORDAN ROSE

TEL AVIV, ISRAEL, 1968. The two-year mission at Tel Aviv University loomed ahead. We were "strangers in a strange land." This was the exotic Middle East. We were fearful and uncomfortable in this foreign place. The university assigned a person to help us find housing. He drove us through the surrounding communities. They looked unfamiliar. We felt disconnected.

Suddenly, America appeared. Down a street that ended in sand dunes, we saw a little bit of home. A small house surrounded by a neat lawn. Bougainvillea covered one wall in bright orange flame. Profusely blooming pink-purple oleander hedges bordered the property. "That's it," the kids cried.

It remained to make the rental arrangements. Our guide took me to the lawyer's office. Idly glancing around, he said, "This lawyer is from Iraq." "How can you tell?" I asked.

"Look at the clients' faces." I looked for some sign of their national origin. There were about ten people waiting their turn. I saw no obvious similarities. Nose shapes varied, hair color varied, and faces were typical of the kaleidoscope of physical traits common in this polyglot nation. What was the distinguishing feature? I looked closer. Nothing. Then he whispered in my ear. "Look at their skin."

I didn't see any tattoos, pimples, or patterns of pigmented beauty marks. Still nothing. Suddenly it hit me—they all had a flowerlike scar on their faces.

He explained, "People get an insect bite. It turns into a pimple just like a mosquito bite. The pimple spreads and becomes a big sore that rots and becomes infected. Then you have to go to the hospital. Adults can get bad symptoms, and their faces can be scarred like these people. If you get infected when you are young, the sore isn't as bad. You can't get the sore again. My family knew this and we put our children outside on warm nights to get bitten. We covered their faces to let the bugs bite in an inconspicuous place, like the legs. You can see that I don't have a scar on my face."

He raised one of his trouser legs. A fist-sized, flowerlike scar was revealed. I thought, "This is the mother of all mosquito bites." Then the epiphany. I dredged up long-past memories of a disfiguring parasitic disease caused by a flagellated protozoan, *Leishmania tropica*. The condition is mildly pathological, with few symptoms other than an enlarged vaccination-like scar.* It is so common in the Middle East that it has many names: Jordan Rose, Jericho boil, and Aleppo boil. Clinically it is known as cutaneous leishmaniasis.

An affair with Jordan Rose is relatively benign—a one-night stand with her causes a lifelong scar. But it is her sisters that you don't want to associate with. The other species of *Leishmania* cause horrible pathological symptoms, often ending in death.

FEMME FATALE

The vectors of leishmania are sand flies of the genus *Phlebotomus* that look like hairy mosquitoes. Unlike most mosquitoes, sand flies can make do with moist, rotting garbage and piles of stinking organic debris for breeding. Mosquitoes cannot survive in the intense heat of dry Middle East desertlike environments, but sand flies hide in slightly moist cracks and crevices of mud and stucco walls in the daytime and, revived by the

---

*In fact this is a form of folk vaccination. It is ancient, predating the work of Jenner and Pasteur by, perhaps, a thousand years.

night's cool breezes, fly off into the darkness to suck the blood of any available host: gerbil, rat, cat, dog—or human. This lack of particularity maintains a reservoir of infected blood. Even if a drug is developed to eradicate the disease and few infected humans are available for the fly to feed on, rodents and domestic animals become reservoir hosts. Inside their sick bodies the infective first stage of the parasite's life cycle malevolently waits for life-sustaining passage into a fly's body in which develop hordes of slender, undulating procreative needles.

In all forms of leishmania, it is the female fly that transfers the infection. As in most invertebrates, the female is larger than the male, the better to sustain the bulk and volume of the eggs she must carry. She needs blood! The energy-rich, protein-rich blood provides the essence of life—all the necessities for egg development. Smaller males bear the tiny burden of sperms and are harmless fruit juice eaters.

The female fly bites a sick human or animal, sucking up infected blood. The parasites reproduce wildly, their vast numbers soon filling the fly's body with spermlike infective swimmers. Some permeate its salivary glands, whose anticoagulant-laden saliva accompanies every bite to keep the blood from clotting quickly. The fly stabs the skin with its stiletto-tipped proboscis. In flows the saliva and with it, hundreds of microscopic invaders.

What next ensues is a horrible perversion of the body's defenses. The attack does not merely bypass the body's Maginot Line. Blood-borne defenders become traitors. White blood cells are tricked into insidiously carrying the alien army into the inner recesses of the body—in some cases right to the heart.

The sand fly bites. Hordes of parasites enter with the saliva. White blood cells rush to the body's defense. They engulf the invaders. But instead of being destroyed, the infective agents remain inside these killer cells and reproduce. They subvert the very white blood cells that congregate to destroy them. The first line of defense is turned against the body in a grotesque distortion. *The white blood cells have made the body more vulnerable to the invaders.* Furthermore, their refuge inside the white blood cells hides the parasites from the ever-threatening antibodies.

In the sixteenth and seventeenth centuries, slaves were carried from the west coast of Africa to Jamaica and other distribution centers, thence to sugar cane plantations all over the Americas. They brought evil with them. Some slaves were infected with leishmaniasis. The parasites were welcomed by local sand flies.

JORDAN ROSES AND ROTTING NOSES

In South America, a different species of the parasite, *Leishmania braziliensis*, descended like a plague from the blood of Africa. The parasite, not satisfied with the relatively benign symptoms of skin-scarring leishmaniasis in the Middle East, added a terrible, torturous twist. It infects mucous membranes. Instead of a petaled scar, *it rots off the nose or ear* in a disgusting distortion of the Jordan Rose.

The fly bites. An itchy pimple appears. This is the primary lesion. In many cases it does not heal, remaining a scab-covered hotbed of parasites. In this muco-cutaneous leishmaniasis, the lesion metastasizes into daughter lesions that fuse together into a rotting carpet of suppurating skin. These infected open wounds cause the majority of deaths by secondary bacterial infection. The infection can spread from inside to the mucous membranes of the nose, a particularly vulnerable area. Soon (or many years) after the initial lesion, a red encrusting, cancerlike deterioration of the inner membranes of the nose or throat causes cartilage or soft tissues to disintegrate. In advanced cases, the nose is just a raw, festering hole in the face. This terrible disfiguring disease is locally called espundia.

THE MOST TERRIBLE OF ALL

It is when the parasite has discovered the richness of the inner depths of the body that it becomes its most virulent form. This variant of leishmania, *L. donovani*, does not get off the subway at the first stop as does its less harmful relatives. It is carried by the body's transport system to the end of the line, the spleen, liver, intestinal wall, and heart.

The spleen is the home of the white blood cells. The spleen is the ultimate filter, the purifier that removes pathogens. White blood cells called fixed macrophages line passageways through which bacteria and other infective agents are carried. They don't float in the blood;

they remain stationary and grasp and grope at invading organisms in a gauntlet of death. This is the body's last line of defense. But the leishmanial parasites are hiding inside mobile white blood cells and are not recognized. They don't trigger the defensive response. So the tiny protozoan parasites have been carried deep into the body, free to invade vital organs.

In the vulnerable liver, the *chemical* filter of the digestive system, the blood flows slowly through tortuous channels. Infective leishmanial parasites enter liver cells and in this dark paradise reproduce, causing massive infections. But they have burst forth from their protective white blood cells. Unmasked, they are vulnerable! White blood cells rush to the region attracted by the smell of death. The spleen and liver swell hugely with their burden of swollen infected cells and massive numbers of white blood cell attackers. A fluid called ascites accumulates. The tragic picture of children with distended, swollen abdomens is a common sight in endemic areas. The disease, kala azar or dum-dum fever, is often fatal. Further, until recently, serological tests were ineffective because the parasites are hidden inside the macrophages. Consequently there was no early warning immune picture to help diagnosis. Kala azar ran rampant until a few years ago, when highly sophisticated immunological techniques were developed to diagnose what appeared to be a disease with symptoms not unlike any other fever-inducing ailment—low fever, running nose, and so on.

DARWINIAN DILEMMA

> *At night I experienced an attack (for it deserves no less a name) of the* BENCHUCA (VINCHUCA), *a species of* REDUVIUS, *the great black bug of the Pampas. It is most disgusting to feel soft wingless insects, about an inch long, crawling over one's body. Before sucking they are quite thin but afterwards they become round and bloated with blood.*
>
> —CHARLES DARWIN

The great mystery in Darwin's life is, what caused him to be chained to his home, rarely venturing far? Had he been more mobile and active, perhaps he would have been able to defend his theories and not depend on his acolyte, Julian Huxley. But as fate would have it, many of his nineteen books were written after his self-imposed isolation. At home he had the time and solitude to further elaborate on his theories of natural selection. Why did he isolate himself? Did he become agoraphobic in middle age? Not likely for a man who traveled the world in a small ship.

Maybe his sedentary life was due not to a psychological condition but a recurrent, anguishing illness?

Charles Darwin, intellectual father of all modern biologists, lived a quiet life at his ancestral home in England. He was wracked with pain, and "scarcely a day went by when he did not have symptoms." The twenty physicians he consulted during his lifetime could not diagnose the mysterious ailment. It has been suggested that Darwin suffered from Chagas' disease.*

In the above-mentioned quote from his diary, Darwin tells of being bitten by the "great black bug of the pampas," a reduviid bug. This insect is not a fly or mosquito but an inch-long true "bug."† In rural South America it hides in thatched roofs during the day and comes out to feed on warm blood at night. Like its relative the bedbug, *Cimex lectularius,* it is attracted to warm skin and carbon dioxide emitted by the human body. The penchant of reduviid bugs for biting the face has led to the common name "kissing bug," but in the case of the South American kissing bug, *Rhodnius prolixus,* and its kin, this may be the kiss of death.

NIGHT TERRORS

Night. In the darkness, hidden inch-long black insects edged with alternating bands of pale rose crawl down the wall in a grim march.

---

*Arguments for the Chagas' disease hypothesis were mainly his gastric and nervous symptoms accompanied by malaise and fatigue, as well as the cause of his death, which was probably chronic cardiac failure.[4]

†The word "bug" represents a specific taxonomic group of insects that has been usurped to mean all insects. True bugs are members of order Hemiptera. They are characterized by sucking mouth parts formed into a beak and wings that are thickened anteriorly.

Seeking mammalian body warmth, a bug mounts the body of a sleeping child, searching for a "hot spot," an exposed place where it can
sink its thick proboscis deeply. The blanket prevents easy access, but
the warmth of the child's exposed face and regular breathing promises bloody sustenance. Once on the face the insect bites. The child
cries out in momentary pain. Still asleep, she scratches her face, inadvertently rubbing the bug's feces into her eye. For as it draws its blood
meal, the evil bug defecates its parasite-laced, deadly feces. Mixed in
with the remains of its previous gory meal are hundreds of flagellated
parasites called trypanosomes. It is not the bite of the insect, but its
feces, rubbed into the wound or any nearby delicate membrane, that
transfers the parasite.

This primitive method of infection is called posterior station, in
contrast to the direct injection by the stiletto-like proboscis of a fly or
mosquito called anterior station. Why is anterior station of more recent evolutionary origin? A method of directly introducing the parasite into the bloodstream is clearly more advanced than relying on the
victim to rub the feces into the wound. Feces dry up after a while, killing the load of death. But a nighttime invasion will almost inevitably
cause the host to rub the parasite-laden feces into the wound or eye.

The child awakens. Her eye hurts. Soon it becomes swollen with
the battle between the feces-borne flagellated invaders and her defenses. If she loses, the eye will be destroyed. In rural Argentina and
Brazil, there is an unusually high proportion of one-eyed children,
a healed-over wound where the infected eye was. In other cases the
insect bites other parts of the body. The site of the bite becomes a sore
called a chagoma, the dreaded sign of the kiss of death.

THE KISS OF DEATH

What causes this dreadful disease? Early in the twentieth century a
physician named Chagas reported on a fever-inducing disease common in Brazil. He had found flagellated protozoans associated with all
cases of the illness. Today we realize that the parasite is a relative of the
leishmanial parasite. Of a related genus, it is called *Trypanosoma (Schizo-
trypanum) cruzi*. So close is the relationship that the trypanosomes go

PLATE 2

A. LESION ON FACE OF GIRL can be 2 inches across, will develop into disfiguring scar but no long-term damage; caused by *Leishmania tropica.* Jordan Rose, Delhi boil

B. MAN'S NOSE ENLARGED signifying long-term infection of mucus membranes. Necrotic lesion extends around his mouth; many cells contain spherical leishmanial cells. Mucocutaneous leishmaniasis is caused by *Leishmania brazilienesis.* Some 90 percent of cases in Bolivia, Peru, Brazil. Espundia, uta

C. BOY'S ABDOMEN DISTENDED by swollen spleen, liver, lymph nodes, and a fluid, ascites. Parasite's spherical cells in heart cells produce fatal visceral leishmaniasis caused by *Leishmania donovani.* Kala azar, dum-dum fever

D. KISSING BUG, *Triatoma* spp. To 1¼ inches; blackish; red bars on sides. Bugs feed eight to fifteen minutes. Common in southern United States. It delays defecation for twenty to thirty minutes and moves off host, thus often-fatal Chagas' disease rare in United States. *Rhodnius prolixus,* brownish with reddish wavy lines on back; to 1¼ inches, is most important vector of *Trypanosoma cruzi.* Transmission via feces rubbed into bite wound.

E. SAND FLY, *Phlebotomus argentipes.* Tiny; to 3/16 inch; resembles hairy mosquito. Female needs blood meal to produce eggs; prefers animals to humans. Introduces parasite under skin where it is picked up by white blood cells and dispersed. Vector of *Leishmania donovani,* cause of visceral leishmaniasis.

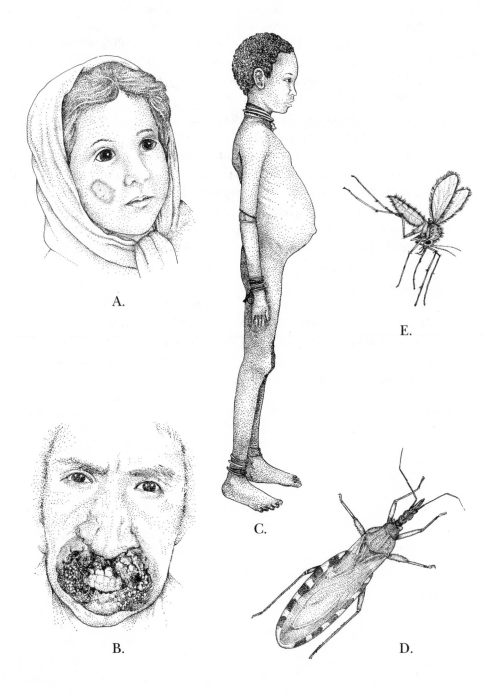

A.

B.

C.

D.

E.

through a leishmania-like stage in their development. Unlike leishmania, they are transmitted by a true bug (not a fly). Unlike leishmania, they are not introduced directly into the blood by a bite. Unlike leishmania, trypanosomes have a stage that swims by means of a flagellum aided by an undulating membrane that propels them through the plasma of the blood. But they do end up infecting not only white blood cells, but heart muscle cells and almost all organs of the body.

Like malarial parasites, they reproduce as spheres that survive in blood cells, not red blood cells like malaria, but defending white blood cells like leishmania. Then they burst forth from the host cell, transformed into the flagellated swimmers called trypomastigotes. The trypomastigotes are eaten by white blood cells, but instead of being killed, they eventually kill their host cell. Since the parasites are protected inside cells like *Leishmania*, it has been difficult to develop a cure for Chagas' disease.

At last count, "twelve to nineteen million people have the disease. 561,000 cases are reported annually. In some surveys in Brazil it has been reported that 30% of all adults die of *T. cruzi* infections. Molecular evidence indicates humans may have been suffering from Chagas' disease for at least 4000 years."[5]

3

I HAD A FARM IN AFRICA*

THE DEPUTY MINISTER OF AGRICULTURE was in a jolly mood. Behind him stood the voluptuous head chemist of the ministry. She appeared to be about thirty years old and was carrying a parasol. Two low-level administrators followed. I was sitting in an old boat with an even older outboard motor—the ministry's research vessel—together with two boatmen. We were moored to a pier on one of the fabled lagunes, elongated lakes that were the estuaries of the many small rivers surrounding Abidjan, the thriving commercial capital city of Côte d'Ivoire (Ivory Coast, West Africa).

Warm greetings turned animated as, with discreetly mild jeers and controlled laughter, the deputy minister and head chemist were unsteadily assisted aboard. The young lady daintily raised her parasol, looking distinctly uncomfortable as the motor sputtered to a feeble, unsteady roar. Off in the distance spread the green-stained waters of the lake, extending about nine kilometers toward the distant sea. The scene reminded me of boating during a summer picnic in the Catskill Mountains back home. This was supposed to be an official field trip during which a site would be determined for the establishment of a huge fish farm. The area that the ministry suggested was near the mouth of the river feeding the lake. The water was shallow and muddy, distinctly unsuited to the deep, semisubmerged net cages that we planned to build.

*Apologies to Isak Dinesen.

The frivolous mood disturbed me. I had not flown twelve thousand miles to participate in a West African picnic. I decided to recommend a lunch break, embarrassingly close to the commencement of our voyage. No one seemed to mind. I treated everyone to lunch of *poulet* and *pommes frites* at a tiny restaurant in a nearby village. We had a good time. Raucous laughter followed every gaffe provided by my college French. After lunch, the symbolic tour was over. The politician/administrators cheerfully departed, leaving the two boatmen. We left the dock again, the motor fitfully propelling the boat down the lake. Using an old French hydrographic map, we took soundings with a rope marked off in meters to which a lead weight was attached. The water was getting deeper. I needed all the depth I could get; the cages were to be three meters deep, and excess food falling through the bottom mesh would rot on the mud below. Decomposition bacteria would use up all the oxygen near the bottom beneath the cages, so I needed about five meters below the cages. Thus the depth had to be at least eight meters.

After puttering down the lake, away from the entry of the river and its attendant shallows, we found an area about ten meters deep in the middle of the lake. Opposite us, about a kilometer away, was a church steeple. I pointed to it. At top speed (as fast as a person can walk) we headed for the church. A small village appeared, surrounded by lush, deep green tropical forest. We pulled up to a flat, grassy area, ideal for a hatchery. A road snaked through uniform rows of oil palm trees and manioc plantations. This was ideal. I said, "This is the place." As we leaped ashore the boatmen excitedly cried, "C'est ca! c'est ca!" (this is it). The site was determined.

Two years later, huge, rectangular concrete tanks covered an area scraped from the forest. The bare red earth looked like a bleeding wound. Bulldozers were still at work. Bare-chested, sweaty men from the village were laboriously filling in the edges of the crudely excavated basins. Nearby a construction truck tilted over at a grotesque angle, its load of gravel spilled on the ground. One wheel was caught in a rain-filled ditch. Bad luck seemed to dog our efforts.

We visited "our" village. According to custom, when a stranger visits, chairs are laid out and the village elders greet their guests. Speeches follow; their meaning clear in any language: "What's new?"

My partner answered. A man of imposing height and girth, he had proven to himself and all doubters that he was a financial wizard. He spoke no French. With impressive self-confidence he flamboyantly emoted, "I bring you greetings from the United States." This was translated into French and subsequently into a mysterious local language. "We are entering a new era of prosperity. The farm will bring employment to the men of the village and we will pay money for their labors." (The village had not yet participated in the money economy and subsisted by bartering.) "We will pay every worker the equivalent of four U.S. dollars a day." A murmur of approval passed through the gathered elders.

The village chief stood. He greeted us with florid enthusiasm and invited us to the obligatory feast of tiny broiled tilapia from the lake. Afterwards, we were honored with a ceremony. The whole village lined up around a clearing, into which a sad-looking goat was dragged by its horns, its eyes rolling and its brown-and-white flanks heaving. It seemed to know its fate. The goat was laid on the ground and, with a rising crescendo of muttering by the crowd, its throat was cut, its blood an offering to the ancestors. The elders, sitting in the front, poured whiskey on the ground as a libation—then drank the dregs. The ceremony was as much for us as to bring good luck to the villagers. A man was almost bitten by a snake the week before, and the aforementioned truck had almost turned over. These were bad omens.

A crowd gathered around a woman with a white-painted face. She was an oracle and had gone behind a tree to consult the ancestors. (A comfortable mixture of ancestor worship and Catholicism was maintained in this village). Her pronouncement was that the project would progress and the bad luck would be dispelled.

All this was taking place not more than twenty years ago and twenty miles away from the skyscrapers and banks of Abidjan, the fiscal powerhouse of West Africa.

THE MOST PREVALENT PARASITE IN THE WORLD

A year later we returned. A sense of responsibility for the villagers had prompted us to bring a young physician. He set up a triage station on the porch of a house and set us to work. The villagers lined up to be examined. My job was to put a thermometer into the right armpit of each person to record their temperature. My fractured French was to no avail as my "patients" spoke Aghien, their own language. The scene was one of good-natured, undisciplined, disorganized hyperactivity. No matter what the diagnosis, everyone asked for pills. It did not matter whether the pill was an antibiotic, an antimalarial, or an aspirin.

Exhausted, I sat down at the end of the day. Our young American doctor walked over to me. He gestured to a child. When the kid came over, the doctor gently held his face and pulled down his eyelid. Instead of pink, the inside of the lid was faded yellow. He called over a little girl. She too had yellow eyelid linings. A dozen children were examined. All had yellow inner eyelids. Based on the sample, it was likely that every child in the village had severe anemia.

The cause of anemia in tropical countries is an often fatal combination of malaria and malnutrition. But it can strike anyone. Even in the professional offices of our Ivoirian partner, with its computer operators, architects, and designers, virtually everyone had to take a week or two off with recurring bouts of malaria, which is as ubiquitous as the common cold, only much more debilitating.

Despite the efforts of modern medicine, malaria still destroys more lives than any other disease except infant diarrhea. But modern medicine is largely the creation of gigantic American pharmaceutical corporations that are required to make a profit for their shareholders. Impoverished nations like those in Africa do not have much clout. Thus, until recently, research into tropical diseases has been placed on the back burner behind antibiotics, antiarthritis medications, and mood regulators. Malaria is very difficult to fight because the infective agent hides inside the red blood cells.*

---

*In 2005 a long-awaited vaccine against the most pathogenic malarial parasite, *P. falciparum*, was developed. It stimulates white blood cells to recognize the merozoites and produce long-lasting antibodies against them. (News in 2008: It hasn't worked in its human trials. Neither have seventy other putative vaccines against malaria reported in 2007.)

The Bill and Melinda Gates-sponsored foundation, PATH, has created the Malaria Vaccine Initiative (MVI) supporting an efflorescence of malaria vaccine research. Their efforts have recently been criticized as being "premature." Estimates as to when the first vaccine will appear start at 2050.[7]

DEATH BY FATAL INJECTION

When a murderer is executed these days, he pays for his crime by fatal injection. But what crime has a child committed that he should be slowly executed by the injection of a parasite? The fatal injection is administered by a most unlikely executioner, a mosquito—not only a mosquito, but a female mosquito, a mother herself. The female of the ubiquitous *Anopheles* mosquito needs to drink energy-rich blood from its victim to provide for her own progeny. If she does not take a blood meal, she cannot produce the capsules of life that perpetuate her species. She sucks up the parasite from an infected person and, weeks later, bites a healthy victim, injecting microscopic infective cells called sporozoites. The sexual act to produce these unicellular progeny occurs in the stomach of the mosquito. These fertilized pre-parasites are then cloned in the mosquito's abdominal cavity in such huge numbers that they invade virtually every nook and cranny of the mosquito's body, spilling over into the huge salivary glands. When it bites, an anticoagulant in the saliva permits the mosquito to sip slowly to satiation. No wonder the saliva is introduced in copious amounts, incidentally carrying its cargo of killers.

The sexual act is the culmination of a parasite's life cycle. Where two hosts are involved, it is the primary host that harbors the sexual phase of the parasite. Thus, because sexual reproduction occurs in the mosquito's stomach, the human is the secondary (intermediate) host of a mosquito parasite. It hurts my feelings to be secondary to the mosquito. If humans are to be involved in a parasite life cycle, is it not fair that we should be the primary hosts? Apparently not.

Upon injection into the human by the mosquito's bite, the sporozoites immediately hide in the victim's liver cells, where they cannot be destroyed by antibodies. Then they metamorphose into infective cells called merozoites that feverishly clone themselves.* After a few days, hordes of merozoites spill out of the liver like invading huns.

In 2008 the first total genome of a human malarial parasite was constructed. This promises to reduce the time for the eradication of malaria.

* More properly called cryptozoites when hiding in the liver.

PLATE 3

A. FEMALE MALARIA MOSQUITO *Anopheles quadrimaculatus*. To ½ inch; must have blood meal to produce eggs. Female is vector of all four kinds of human malaria. Six mouthparts form an enclosed tube, the proboscis (longest structure projecting downward from head). When mosquito bites, sensory structure in proboscis finds a blood vessel. As blood flows up proboscis, parasite-laden saliva flows down. When feeding, body, rear legs point upward at 45-degree angle.

B. SEXUAL REPRODUCTION OCCURS IN STOMACH. Mosquito bites sick human; ingests blood containing male gametocyte (*left*) and female gametocyte (*right*); long flagella of male gametocyte become sperms and swim to female gametocyte, fertilizing it.

Fertilized egg produces many embryos (ookinetes (D)), that burrow through stomach wall, bulging to body cavity; form spherical oocysts (C).

Massive asexual reproduction in oocysts produces many slender sporozoites (E) that migrate to salivary glands of mosquito.

C. MOSQUITO STOMACH with oocysts on outside wall.

D. OOKINETES burrow through stomach wall to become oocysts that produce sporozoites.

E. SPOROZOITES enter salivary glands and are injected when mosquito bites. These infective cells enter human host's blood and migrate quickly to human's liver. This ends mosquito phase of life cycle.

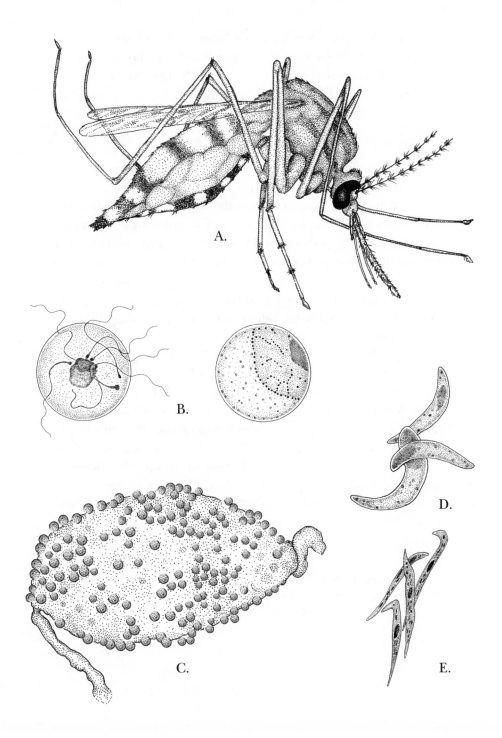

A.

B.

C.

D.

E.

The battle is joined. Many are killed by antibodies in the blood, but a few breach the defenses and enter red blood cells. Protected inside the cells, they reproduce unfettered, eating the hemoglobin.

After two or three days of dividing inside the red blood cells, eight or sixteen merozoites are disgorged, tearing apart the now depleted cell. The massive destruction of red blood cells and the invasion of cloned merozoites cause the body to resist in primordial fashion. For uncountable millennia, the body's defense to overcome any infection was to create an inhospitable environment. First the signal that the body is being overwhelmed reaches the hypothalamus of the brain. It responds by creating the response to cold—prolonged rapid muscular contractions, or shivering. These create heat that is not allowed to dissipate. This sequence causes the body to alternately feel terribly cold followed by very high fever, in cycles of two or three days. But the body cannot create such inhospitable dangerous conditions for long. A fever of 106 degrees Fahrenheit cannot be sustained without harming the brain.

Malarial parasites are living organisms. They can mutate. Four modern species have evolved to be resistant to the body's primitive defenses. The evolutionary battle has been fought to a draw of sorts in three of these species. The periodic debilitating episodes of shivering and high fever, often accompanied by a severe headache (together called a paroxysm), do not usually cause death. The fourth species, *Plasmodium falciparum*, can kill. Even if infected by one of the three less-than-fatal species, many of the children in our village were doomed, the survivors destined to function below their potential because of severe malaria-induced anemia.

In the past few years there has been a worldwide race among pharmaceutical companies and research institutions to produce an extremely inexpensive antimalarial drug or vaccine to allow children in the tropics to reach their full potential.

4

DEATH OF A MOUSE

SOMEWHERE IN AN AFRICAN JUNGLE, millennia ago, a mosquito sucked up a minute droplet of blood from an African tree rat, *Thamnomys rutilans*. In its blood were millions of red blood cells infected with a precursor of a modern malaria parasite.

Some time after that, the mosquito took a blood meal from another rat, salivating madly over its rich red repast. The saliva, thick with the offspring of the malaria parasites, poured into the wound. It contained a clot-busting anticoagulant that kept the blood flowing during the few moments required for the mosquito to pump enough blood to fill the hugely distensible bags (gastric cecae) attached to its stomach. Sated, swollen with blood, the mosquito could hardly fly, but it survived.

This accidental introduction happened over and over again innumerable times.

Having harnessed the omnipresent flying hypodermic needle (the female of the common Anopheles mosquito), the parasite used her to probe different species of reptiles, birds, and mammals. Mutations occurred at a relatively fast rate and new species arose to be parasites of a variety of hosts.

But over time, the rats mutated too. The mutations made them invulnerable to the malaria. Parasite and host came to a truce: a modest number of malarial cells would survive in the rat's blood for a few days, after which the powerful immune system would wipe them out. The rats had become immune to their special malarial parasite, *Plasmodium berghei.*

SEX IN THE STOMACH

What is the malarial monster like? Fangs and claws and scaly skin?

When stained and viewed under the microscope, all malarial parasites are innocuous-looking minute spheres inside red blood cells. But some of these uniform-looking dots are different. Sex is important: it accounts for the potential of the species to change. Under high magnification some of the dots resemble sausages containing bands or crescents. These are the boys and girls that will mature into potent males and fertile females after being sucked up by the mosquito. In an orgy of sexual reproduction in the dark confines of the insect's stomach, male dot finds female dot and reproduction happens.

The male dot, in true masculine style, becomes hairy inside the gut. The "hair" is massed microgametes—sperms. They penetrate the female reproductive cell (macrogamete). Sex, *Plasmodium*-style, has occurred inside the mosquito's stomach. The progeny of these sexual encounters, elongate banana-shaped offspring, penetrate the gut wall and asexually produce thousands of slender, spindle-shaped cells that permeate the mosquito's body and invade its salivary glands. These spindles are the infective phase that the mosquito injects into the new host as it takes its blood meal.

How do researchers, in their efforts to develop antimalarial drugs, attempt to destroy each physically and immunologically different stage? How can the vicious cycle of the host-parasite relationship be interrupted?

The mosquito seems to be the weak link in the malaria life cycle. We can spray half the world with DDT, the most effective mosquito killer. But DDT destroys beneficial insects, inhibits egg development in birds, and otherwise harms the environment. What to do? Formerly DDT was sprayed ubiquitously and ferociously on the assumption that saving lives was worth the damage. Now, in the absence of inexpensive medication, the solution is to limit the harm by spraying the walls of houses so as to minimize collateral damage. This technique is risky, but the alternative, death by malaria, is unacceptable.

Dangerous solutions are not real solutions. Hopefully, the acronym DDT will be replaced by the influence of another acronym, DNA.

Recently, DNA profiles of the confusing maze of life stages of the malarial parasite were unraveled and new vaccines were developed. Millions of doses, in experimental form, were rushed to Africa.* There is finally hope for the three hundred million human victims of the devastating disease, *Earth's number one killer.*

A GOOD-LOOKING MODEL

How were the vaccines created? The tree rat eventually evolved a resistance to its parasite. Not so the common house mouse, *Mus musculus.* Closely related to rats, mice are vulnerable. Having evolved in a far different habitat from that of African trees, the mouse shared the rat's rodent genes, but not the few "new" genes for resisting the malarial parasite. Vulnerable, the mouse dies while its kin survives.

Four variants of the ancestral human malarial parasite were known to exist before the war in Vietnam. There, sick soldiers exhibited malarial symptoms, but the disease would not respond to traditional antimalarial medications. After a hysterical search, the existence of a new species of malarial parasite, *Plasmodium mekongae,* was postulated and named after the battleground—the Mekong River delta. This new malarial variant was as dangerous as the Vietcong enemy.[†]

How to study this new and dangerous parasite? Researchers could not try out new drugs on sick soldiers—that would further endanger them and is unethical. No, a model for studying the *Plasmodium* parasite would have to be found. A few years before, a curious researcher was abstractly studying a then-obscure malarial parasite of tree rats— *Plasmodium berghei.* He injected a mouse with malaria-laced

---

*Among the other promising new methods for inexpensively inducing immunity involves mosquito spit. Saliva from uninfected mosquitoes seems to stimulate the production of antibodies (cytokines) against malaria. How does one collect enough spit?

[†]Subsequent research revealed that the putative *P. mekongae* was in reality a drug-resistant form of *P. falciparum,* the most dangerous of the human malaria parasites.

PLATE 4

A. MOUSE EXPERIMENT. Each pair of students received two white mice bred exclusively for research. When experiment began, red blood cell (rbc) count of both mice was over eight million cells per mm³ of blood. Both were alert, active.

Both mice weighed on day 1; blood count taken.

Experimental mouse injected with millions of malaria-infected rbcs.

After 14 days control mouse remained active; weight stable. Blood count remained the same. Experimental mouse's (*foreground*) rbc count declined after day 3 from eight million to two million on day 14; shivered uncontrollably; somnolent; fur matted, infused with urine; eyes slits; nose pale; too weak to eat, lost 30 percent of weight. Died on day 14.

B. MAMMALIAN LIFE CYCLE IN MOUSE

    (1) INFECTED LIVER CELL. Sporozoites (plate 3) injected into blood, enter liver cells; divide into many merozoites that spill into blood.

    (2) RING TROPHOZOITE. Merozoites enter red blood cells; become trophozoites; feed on cell. Nucleus stains ruby red—called "ring troph."

    (3) SCHIZOGONY. Merozoite divides asexually; becomes schizont; eats hemoglobin in host cell; becomes eight or sixteen merozoites.

    (4) RED BLOOD CELL BURSTS releasing merozoites and wastes. Most killed; few enter rbcs. Wastes influence brain to produce paroxysms.

    (5), (6) GAMETOCYTES APPEAR IN A FEW RBCs. Female (*lower*); male (*upper*) gametocytes ingested by mosquitos beginning insect phase (plate 3).

    (7) UNINFECTED RED BLOOD CELL. Mature rbc has no nucleus.

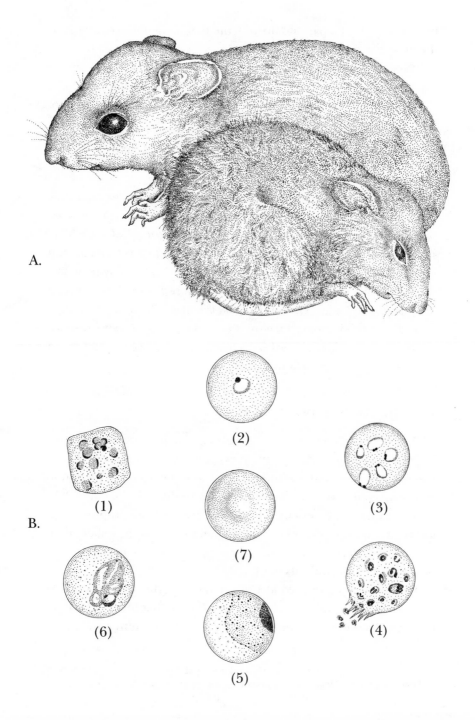

A.

B.

(1) (2) (3)

(4) (5) (6) (7)

tree rat blood. The parasite thrived; the mouse died. Here, then, was the model!

The mouse was extremely susceptible. It died in two weeks, rapidly expiring from the same symptoms exhibited by the soldiers.

Much of the initial research was done at New York University School of Medicine nearby.* It struck me that *P. berghei* was an ideal organism for teaching about malaria. Many of the students in my parasitology class were premeds and potential researchers. What better preparation could they have for their future careers?

NYU lab-infected hamsters were royally transferred by car to my lab. They were originally infected with blood from a tree rat, and the parasite was passed down over the years by injection. Infected hamsters became a parasite-rich source of *P. berghei* in lieu of tree rats. But who knew when the malaria would kill them? Immediately their tiny tails were punctured and drops of blood were injected into several white laboratory mice. Only then could we relax. The white mice were to be the reservoir from which the students would spread the disease, taking on the role of the hordes of mosquitoes that decimated African villages—or army barracks in Vietnam.

TRYING TO SAVE DOOMED MICE

Each pair of students received two sibling female mice. One was injected with blood infected with *P. berghei*; the other wasn't. (Fortunately it was possible to transfer the parasites by injecting them into the peritoneal cavity of the mouse—the students lacked the skills to find a tiny blood vessel.) After two or three days a drop of blood was removed from the tail of the mice and smeared on a slide, then stained purple. A tiny glass straw was used to suck another drop of blood for a red blood cell count.

The number of red blood cells of the experimental mouse was beginning to decline. The students were aware that little "Mary" was doomed. They were determined to save her. They scoured the litera-

---

*I here acknowledge with appreciation the help of a pioneer in *P. berghei* human malaria research, Dr. Meier Yoeli, in providing wise counsel and infected mice.

ture desperately. Each pair chose whatever potential salvation seemed to show promise. Discovering the famous relationship between gin- and tonic-swigging British expatriates and a lower incidence of malaria, some students decided to replace the mouse drinking water with tonic (= quinine) water. Others just tried the gin, reasoning that if the mouse died at least it would die happy. Some pairs of students decided to add or remove from the diet nutrients associated with growth or reproduction of malarial parasites. No matter what they did, their graphs of red blood cells showed a steep decline compared with the consistently straight line of the control mouse's blood. The tragedy was about to unfold. After two weeks the experimental mouse was somnolent and shivering in the corner of the cage. The red blood cell curve had reached its nadir. The mouse, warm and curious and healthy two weeks before, was dying. The students were inconsolable. The drama of life and death had played out.

No need for me to explain what effect this real-life experience had on the intellectual development of each future researcher.*

POSTSCRIPT

An article in *New Scientist* (A. Coghlan, May 26, 2007, p. 16) described a new "Super Antibody" discovered in 2007. "Antibodies taken from Gambian people who are immune to malaria could be used to protect others from infection. The researchers started by 'humanizing' a malarial parasite that normally infects mice. . . . to do this they genetically modified *Plasmodium berghei*."

I like to conjecture that one of my former students might have been a researcher in this group.

---

*A few students learned that they had no stomach for research with living animals and changed majors or took more theoretical biology courses such as DNA genetics, where the animals that contributed the DNA are not visible.

5

INTIMATE RELATIONSHIPS

THE VOICE ON THE TELEPHONE was unmistakable. The imperious tones of the boss projected the omnipotence of royalty. Never mind that the ancestors of Alexandra Proskoriakoff-Minsky were probably thrown out of Russia with the rest of the Jews. The fact that her regal voice came from on high (Jerusalem) added to the feeling that one must obey her command. "A team of British experts is coming to Israel for a visit and we need you to entertain them. Show them around the Old City (Jerusalem)." "I am very busy, Alex," I protested. "None of us has the time, and we can't speak English that well," she retorted. (Everyone spoke excellent English.) I lost the argument.

The next day I met the British team, a jolly bunch. The tour started at the imposing ancient Jaffa Gate. David's tower brooded over high crenellated walls made of chiseled blocks of golden limestone. The bright sun highlighted shadows and hoary black spots that may have been lichen—or the blood of ancient pilgrims. I took my gaggle of professors to the dark, narrow streets of the venerable souk. Their rowdiness disappeared as we walked among kaffiyeh- and skullcap-covered heads. The presence of the deities of three religions was palpable.

The streets were crowded with pilgrims from every corner of the world, jostling my proper British guests as we made our way through the winding streets, almost overwhelmed by the pungent smells from spice stalls. A butcher shop where dead, fly- covered goats hung from hooks was pressed against its neighbor, a vegetable stall where every tomato gleamed with a hand-polished glow. The sun rarely penetrated. When it did, it lit up the colorfully embroidered black dresses of passing women, their dark veils adding a note of mystery.

When the heat of midday brought beads of sweat to my charges, I led them under the arched gateways away from the human hordes and darkness of the souk. A right turn brought us to a dingy, narrow street. It was a cool, peaceful contrast to the bustling, pungent market. The restaurant was at the end of the street. There was an incomprehensible sign in Greek over the door. The menu was translated from the Greek into equally incomprehensible Arabic, but I knew that the pièce de résistance was stuffed pigeon (not squab in our Western vernacular, but what it was, a pigeon). It was identifiable as such when the dishes arrived with the head still on the bird. There is a technique for dealing with the dead bird's staring, accusatory eye. I showed my guests how to eat the pigeon without guilt. The dish is served with "cheeps." (Hebrew does not have a soft "i," so the word for potato chips has been distorted. A further distortion transformed the word to mean French-fried potatoes, as in "fish and chips.") The technique is to pile the "cheeps" on the head, obscuring the glaring eye. It then becomes possible to tear away at the abdomen, exposing the wild rice and pine nut stuffing, ignoring the hidden head.

In the Middle East, hospitality requires that a meal must be preceded with a minifeast of appetizers consisting of salads. Humus (mashed chick peas) forms a ring around the edge of a plate, surrounding tahini, a thick, milky fluid made of ground sesame seeds. One dips a chunk of pita (bread) into a communal dish, swiping up some of the humus/tahini mixture. I felt that it was necessary to further overwhelm my guests with the ultimate sumptuous feast. I ordered all the salads. This excess of hospitality included parsley salad. That was my undoing. Even in this restaurant, with its immaculate tile walls and floor, danger lurked.

It is the custom in poor countries for farmers to fertilize their crops with the eminently available human feces. If the chef does not wash the vegetables assiduously—disaster! The parsley must have been raised by a very sick farmer. A day or two later I awakened with pains in my joints, diarrhea, and a splitting headache. I couldn't get out of bed except to rush to the toilet. Weakly, I called the leader of the British group, fearing the worst. "No problem so far," said he. I railed at my bad luck. Of all the people in the group, I was the only one to get sick. Then I remembered that the cause of the hilarity was in a flask passed around the group. The British have evolved a technique for dealing with the dangers of the Third World—drink gin. The ubiquitous gin and tonic is the expatriate's salvation. Causing the bitterness of the "tonic" is its major ingredient, quinine, the traditional antimalarial medication (it provides protection against malaria, but the alcohol probably suffices to protect against all other parasites).

A day later, I was able to stagger into the local physician's office. I stood at the door, weakly holding on to the frame. "Dr. Sugar (that was his real name), I'm sick." He took one look at me and said, in his Viennese accent, "Ven vas you in der Old City?" "How did you know?" I responded. He looked owlishly at me and prescribed an antibiotic. "You have bacterial dysentery. The pain in your joints comes from bacterial toxins producing a temporary arthritis-like condition." The symptoms disappeared in a few days. Little did I know that my intestinal encounter was not over.

THE DAY I GAVE BIRTH

About six months after my return home, I incidentally looked into the toilet after using it and mused, "My, there is an eight-inch-long worm in there." It took two thought processes to realize that *I* had "given birth" to the worm, and that it was the infamous pencil-thick roundworm, *Ascaris lumbricoides*. I had ascariasis! I was almost pleased when I realized that I had acquired one of the most common parasitic infections in the world. I would save the worm and show it to my parasitology classes. I would explain that it was "*my* worm." This would emphasize the ubiquity of parasites. (Years later I lost the worm on a bet. I had wagered my most precious possession on a sure thing—oh, the foolishness of gambling!)

I passed no other ascarids. Evidently, this was a rarely found single-worm infection. It was not a surprise that I felt no symptoms. Many parasites produce no noticeable effects. It is not advantageous for parasites to cause harm to their host. If the host is impaired or dies, the parasite dies. In general, the more highly evolved the relationship, the less harm. The environment of the parasite is the body of the host. If it has had time to evolve over the millions of years of human existence, the worm will become intimately adapted to the physiological and immunological parameters of its host, evolving over time toward the host's ultimate survival.

DEATH OF A CHILD

The only occasion of pathology caused by *Ascaris* I know of personally was related to me by my dermatologist. In her youth she had worked in an impoverished country. A child was brought to her, writhing with a bulging belly and much pain. Lacking sophisticated equipment, she was unable to diagnose the condition and the child died. An autopsy revealed that the child had so many ascarid worms in her gut that they blocked the dribble of food that was available to her. She literally starved to death. Tears welled in the eyes of the physician as she recounted the story. "Medical schools don't usually offer courses in parasitology, and I was untrained to look for symptoms of parasitic infection."

Few healthy people die of ascariasis. But the impoverished are vulnerable. It is ironic that while the starving can host a squirming mass of worms, the wealthy and healthy are relatively invulnerable, with far fewer serious infections. Although most fecal contaminants contain hundreds of eggs, only a few adult worms are found in the strong host's intestines. There is evidence that the environment regulates population size. In tiny worms oxygen can diffuse through the body wall. Although tiny, *Ascaris* juveniles will not develop past a certain stage at low levels of oxygen. When they become larger they cannot obtain more than 50 percent of their requirements through diffusion. There is a low concentration of oxygen in the gut, primarily along the walls. Thus oxygen availability becomes a limiting factor in *Ascaris* populations. Nevertheless, the weak and impoverished manage to defy the odds and become hosts of gut-plugging masses of worms.

PLATE 5

A. INDIA. PUBLIC HEALTH TEAM arrived in impoverished hamlet and administered anthelmintic to population. A few days later this foot-high pile of *Ascaris* worms had accumulated. Many multiple worm infections, especially among children. (Picture adapted from UNESCO photo.)

B. *Ascaris lumbricoides.* Adult female (upper) white, pencil-thick, 8–14 inches; male 6–12 inches; 1/4 inch thick, hooked tail. In temperate and tropical zones. Poor villages can have 100 percent infection; some people harbor more than one hundred worms. If worms inside all humans were placed head-to-tail, they would encircle the world fifty times.
Migrating juveniles can irritate lungs and cause asthma attacks. Most adults don't live more than a year, but successive exposure causes permanent infections. Sometimes, if disturbed, adults can migrate out of nose, mouth, or anus, causing severe emotional distress.

C. SMALL INTESTINE DISSECTED OPEN to reveal many *Ascaris.* Sometimes worms clog intestines, causing death.

D. *Ascaris* EGG needs two to three weeks outside host to become infective. Thick, bumpy shell prevents drying out; eggs can survive for months on plants or in contaminated drinking water. Each female lays 200,000 eggs per day; 73 million during her lifetime. Estimated weight of *Ascaris* eggs in China alone is 18,000 tons.

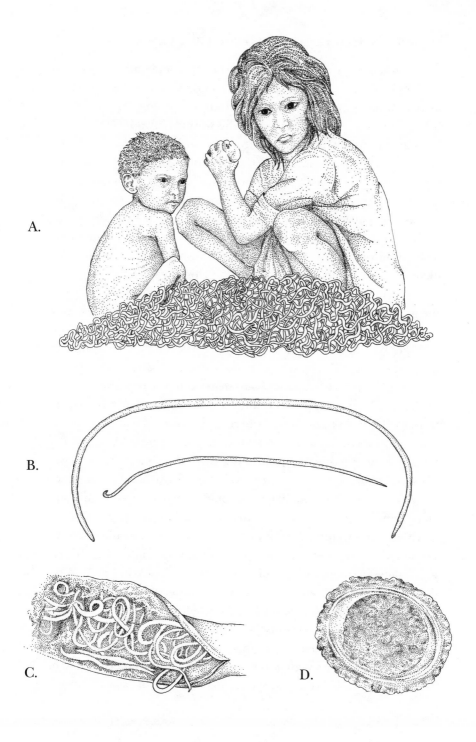

A.

B.

C.

D.

HOW TO FERTILIZE EGGS WITHOUT A PENIS

Ascaris lumbricoides is the human version of a ubiquitous family of roundworms that infects virtually all vertebrates. *Toxocara canis* infects dogs, cats, and other carnivores. *T. cati* has similar hosts; *Ascaridia galli* infects chickens. The insidious way the worm is introduced into the body is responsible for its near-universal distribution. The egg is covered with a thick mammilated (bumpy) shell, allowing it to survive long after the feces have rotted away. Microscopic, it is not noticeable on raw vegetables that have been fertilized with human excrement. In my case the parsley was unwashed. *I ate feces!*

The strong acid of the gastric juice acts as a gate, refusing entry to harmful organisms. But parasitic roundworms have had millions of years to solve the conundrum of the closed gate. After the thick shell has repelled the turbulent gastric defenses and the egg enters the intestine, the juvenile worm inside its shell haven detects a barely perceptible cue. Shell-cracking secretions are produced, freeing the juvenile. It escapes into the all-providing intestine. But like a childish movie star, it is too immature to handle all this wealth. It must get to the parasite nursery—the lungs—to mature. It burrows into the wall of the duodenum (upper intestine) and seeks the rich red freeway through the heart to the lungs, where it penetrates the air spaces and molts to an advanced stage. After about ten days, upward-beating cilia lining the lung passages and pharynx carry the almost-mature juveniles to the throat where they are swallowed. Some are too immature to survive life's rigors and, like those childish movie stars, commit suicide. They plunge into the acidic hell of the stomach and die. But the more mature juveniles have grown a cuticular coat, invulnerable to the stomach acid, and survive. There, in the upper intestine, wallowing in digested food, the juvenile becomes an adult. Within sixty days the females have grown to their usual eight-to-ten-inch size and are ready to be fertilized by the smaller hook-tailed males.

But the males have no penis! What to do? In an unusual solution to this procreative problem, they insert two needlelike copulatory spicules into the female's vagina. But spicules can't move spermy semen. The solution: evolve ameboid sperms to crawl over the spicule-bridge in an amorphous parade of male genes. The shapeless sperms crawl up the uterus and wait. As the eggs are released by the ovary, sperm penetration occurs. The cycle is complete.

This ghastly story of unsanitary eating still makes me shudder. It is not an abstract lesson. *This happened to me.*

6

A PEEK INTO THE ANUS OF—
MY CHILD

PITIFUL CRIES EMANATED from three-year-old Julie's room. She was wailing, her eyes closed, half asleep. "My vagina hurts." We could find no cause—no redness, no swelling. "Go to sleep, baby," I crooned. The next night, the sad scene repeated itself. We brought her to the pediatrician. The doctor thought a few moments and pedantically made a pronouncement. "She has a pinworm infection." "Impossible," I blurted, repeating in my mind the notes I had recently taken in a course in human parasitology at the medical school. "First of all, the worm crawls around the anus, not the vagina. "Second, it causes itching, not pain." The pediatrician's comment was not just a diagnosis, it was a challenge. "How could I not have identified so common a problem as a pinworm infection? I have just finished breaking my brain over innumerable life cycles of every human parasite. Is it possible that I failed to diagnose the most common childhood parasitemia?"

The kindly little woman looked at me with a steely glance, obviously implying that I should not be so audacious as to doubt her diagnosis. She abruptly carried my daughter to the examination table, flopped her over, and with two fingers opened her anus. Swinging over a bright surgical light suspended from the ceiling, she said, "*There it is.*"

In the intense glare of the light, I could see a half-inch-long whitish worm sinuously moving against the pink background. "So much for theory," I thought.

CAVORTING IN THE ANUS

The roundworm we call the pinworm, *Enterobius vermicularis*, is unique in that its eggs are unusually tiny motes of nascent life, so tiny that they can float in the air as lightly as dust, and thousands can fit under the fingernails of a child. The bizarre life cycle requires a child–worm connivance. Each egg contains a developing larva—little more than a string of cells encapsulated in a glassy shell. The massive number of eggs under the child's fingernails inveigle the innocent child into participating in the ritual of infection.

The female worm meets the male in the colon; they mate. He dies and passes out in the feces. The female, bloated with thousands of glass-walled eggs, eventually ends up in the rectum. She is not attached to the wall. She is mobile. Every night (how does she know it is night?) when the child is fast asleep, she creeps out of the anus and slithers onto the perianal area. And slither she does, for roundworms lack the layer of circular muscles needed to create ripples. No ripples, no purchase on the skin, hence labored forward movement. She thrashes wildly, making little progress. This causes a very itchy sensation (not pain—when a child is awakened from sleep, the two sensations seem similar).

To complicate the issue, uterine secretions released during egg laying irritate the delicate skin.

The intense itching requires a response. The sleepy child unwittingly accepts the worm's evil invitation—she scratches the itch. Under her nails lies the potential to infect every child in her class. The next day, unaware of the drama that unfolded in her perianal area, she goes to school. Teacher sees to it that she wipes her hands after washing—on the class towel! Thousands of infective eggs are shed from under her fingernails.

She returns home after school—the cutest carrier of disease, the cutest wormy collaborator—shedding eggs everywhere she goes. In addition to infection via contaminated food or towels, mom and dad can breathe in the eggs, as can sis and bro.

Thus does the pinworm spread from classroom to classroom, from place to place, from country to country. It is ubiquitous. It is not associated with poverty. It can occur in the middle of suburbia. In some countries adults have a lower incidence of infection than children, suggesting that some sort of immunity is conveyed after repeated infections. In these countries baths are uncommon and clothes go unwashed. But there is no guarantee that any child will avoid infection, so universal is this parasite.

Although the infection causes little harm, the hysteria it evokes begs for a solution. In the old days, forty years ago, my wife was faced with an almost insurmountable problem once the disease was diagnosed. The only solution was to boil the sheets, the towels, the bedclothes— of everyone in the house. One soiled towel could cause reinfection. Nowadays the drama is out of this disease. So-called miracle drugs like albendazole/mebendazole will eliminate the infection from the host, but the boiling must go on to prevent another occurrence.

PARADISE

Paradise, for a roundworm, is life in the bloodstream or intestines, bathed in warm fluid containing every necessity. There is enough oxygen along the walls of the intestines to sustain a roundworm's slow metabolism. Until recently it was thought that some species that eat red blood cells, like hookworms, get oxygen while munching on the hemoglobin. Not true in the light of modern knowledge. *Ascaris*, like the other roundworms, is content with digested food molecules gulped down by its three-lipped mouth. Or, as in the case of our pinworms, abundant intestinal bacteria will be the first course, topped off by a dessert of epithelial cells lining the gut wall.

But Paradise is not easily achievable. Apparently, overwhelming barriers lie in wait: first the fiery pit of the stomach, its corrosive acid a formidable barrier. Then the proteolytic enzymes of the intestinal juices, which nibble away at the proteinaceous body of the parasite. The road to Paradise is bifurcated by two evolutionary paths. One leads to death, the other to riches never thought of by humans—the opulence of feces; the super-charged oxygen-rich blood.

HOW TO GET TO PARADISE

There are four universal methods that parasitic nematodes use to break into the blood and guts of the human body. A fifth variant, that of pinworms, is discussed separately below.

The first way is to meet the problem head-on. Food contaminated by egg-laden feces is swallowed by the host. Once the eggs pass through the stomach acid, the second half of the host's one-two punch must be dealt with. The deadly dissolvers, the intestinal proteolytic enzymes, are the next gauntlet that must be passed. What solution has round-worm evolution provided? The same solution much of our world depends on—plastic. We live in a plastic world. This book is on a plastic table, you are holding a plastic pen and reading with plastic eyeglasses. Plastics industries form the foundation of some national economies. But roundworms (hereafter called by their scientific name, nema-todes) invented a plasticlike cuticle countless millennia ago. When the egg hatches and the juvenile worm emerges, its "plastic" armor is impervious to the enzymes that are digesting proteins all around it!

This sort of "direct" life cycle, where the egg is ingested and hatches in the gut to eventually produce another adult that lays eggs, and so on, is exhibited by *Ascaris*.

The second mode of entry, used by hookworms and their kin, starts with the release of five thousand to ten thousand eggs per female per day in the feces. Inside each egg are four or eight cells destined to become three-lipped mini bloodsuckers. They mature into juveniles that burst out of their eggs in freshly deposited feces. Most die, dried out by the sun's searing rays. But if the eggs hatch in moist, warm areas, such as the humid shade under a tree, they enter into a free-living stage, eating rotting detritus, writhing through the rot long after any sign of their fecal source has disappeared. Occasionally a few of the free-living juveniles morph into a nonfeeding infective stage.

A barefoot child wanders along. The organic soil mushes under his feet and squishes up between his toes, carrying these infective juveniles with it. They promptly burrow into the soft skin between the toes and follow a long, preprogrammed path from skin capillaries through heart and lungs into the digestive tract, where they grasp the walls with a disproportionately huge mouth and chew their way into a capillary-rich layer, feeding thereafter on red blood cells. Massive infections of these hookworms, *Necator americanus*, can remove two cupfuls of blood a day, causing hookworm anemia. The image of a somnolent hillbilly sitting on his porch with a jug of homemade moonshine is misleading. He is not drunk. He suffers from hookworm anemia. Just look at his shoeless feet.

A VERY DEDICATED SCIENTIST

The first description of hookworm development was made in 1896 by an unusually dedicated parasitologist named Arthur Looss. Deep in his investigation of hookworm disease in Egypt, he was feeding hookworm cultures to guinea pigs. Some of the living worms spilled on his hand. Curious, he didn't wipe the worms off. After a few minutes, he noticed that the site became itchy and red, suggesting that the infection is spread by skin penetration. Sure enough, after a few weeks, his stool contained many hookworm eggs.

To prove his point, he found an Egyptian boy whose leg was about to be amputated. He poured some of the same worm culture onto the boy's irretrievably injured leg. After the operation he held the severed leg in his hand. While it was still warm, he cut off chunks of flesh and processed them for study. Under the microscope, paper-thin sections of the still-living skin showed the juvenile worms burrowing into the peripheral blood vessels. A description of the process appeared in his famous paper.[8]

A PIG EATS SLOP

A third mode of entry is to form "cysts" (actually, capsules) in the muscles of an animal, e.g., pigs, bears, and porcupines. No eggs are involved. This is live birth, just like humans. If transmission does not require an external egg, how can a parasite in a pig be transmitted?

Simple. Eat the flesh of the pig. Buried inside the pork are hundreds or thousands of juvenile worms, each waiting for your help in freeing it from its self-constructed prison.

1. A pig eats "slop." In this disgusting mélange of filthy food is a dead rat (the farmer smiles as he thinks "all the more protein for my fat friend").* The rat had feasted on a newly dead pig, bear, mouse, or, rarely, flesh-eating beetle and had become infected.

As the pig munches on the hairy rotten lump in its food, the swallowed chunks of infected tissue pass into its gut, freeing the juveniles to continue the life cycle. After a month or so, the pig's tissues are riddled with "encysted" worms.

2. You like your meat rare. Your host obliges, serving up a barbequed pork chop, pink glowing enticingly from each slice. Soon after, your digestive juices free the juveniles coiled up in their version of a fetal position.

3. These juveniles burrow into the gut wall to mature and mate. The pregnant females thread their way through the gut wall, entering blood vessels. They give birth to live young that migrate through a bloody pathway: liver, heart, lung, circulatory system—spreading throughout the body, preferentially entering muscle cells and subverting them to their own needs.†

4. Somehow the juvenile alters the very essence of the cell (the expression of its genes) and changes the cell to be a nursery. Eventually, in a perverse collaboration, parasite and host together shroud this "nurse cell" containing the juvenile into a scarcely visible white, calcareous, cystlike capsule. There it waits to be eaten by another host. In humans it is doomed to wait to eternity.

Thus *Trichinella* has the dubious distinction of being the largest intracellular parasite known.

---

*Nowadays, huge, extremely sanitary factory pig farms produce virtually all the pork we eat. But a few small farmers still produce pork the old-fashioned way for local consumption.

†In the process of invading the muscle cells, the juvenile worms lyse (dissolve) the muscle fibers, weakening arm and leg muscles. But for unexplained reasons, most juveniles accumulate in the jaw, tongue, and diaphragm. It is these muscles that are examined by meat inspectors.

This particular worm, *Trichinella spiralis*, can be considered a pig (or bear) parasite, normally transmitted by cannibalism, that accidentally has been introduced into humans. It is dangerous. In severe trichinosis there can be intense muscular pain, difficulty in breathing because of diaphragm muscle inflammation, heart damage, and nervous system involvement including hallucinations as the juveniles wander through the body. Death might be caused by heart failure, respiratory complications, and impaired kidney function caused by poisons released by the worm's metabolic wastes—or a massive immune response to the migrating juveniles in the blood.

A story is told about the theft of some newly made pork sausages riddled with cysts. All members of the thief's family died of trichinosis after eating the undercooked sausages, except the mother who did not like pork.

These three methods of infection rely on millions-to-one chance encounters between parasite and host. A fourth method, more recently evolved and highly sophisticated, involves the direct injection of the parasite into the host. Not only is the element of chance eliminated, but the paradise found is the richest of all, the blood.

A mosquito sucks blood containing juvenile worms from an infected person. They mature inside the mosquito and, when the mosquito takes a subsequent blood meal, are released onto the skin of a healthy host. (They are not injected as are the malarial parasites; they crawl into the bite wound.) From Malaysia to Borneo and across the equatorial world, the microscopic juveniles of the roundworm *Wuchereria bancrofti* mature to cause that grotesque condition, elephantiasis.

THE ODD PINWORM

The object of our infection, the pinworm, has been discovered in fossilized feces (coprolites), indicating that the worm has coevolved with humans since their appearance on the evolutionary scene. Given all this time, the pinworm has evolved yet another, more effective style of transferring its young to your young. Its eggs do not pass out in the feces to be subject to the elements. The female worm entrusts them to your child's care. She migrates out of the anus at night, thrashing vigorously, wandering afar in the perianal region, strewing dust-sized, glassy-shelled eggs in her sticky path. She slithers far in her murky world, sometimes reaching the child's vulva, depositing up to sixteen thousand eggs every day. Sometimes the females are so pregnant they explode.

A few eggs hatch after deposition in the perianal area, releasing infective juveniles. Squirming as if they know their life depended on it, they seem to be trying to reach their anal haven and crawl back to their ancestral breeding place, like salmon going upstream. This method, "retroinfection," is uncommon.*

The usual transference procedure is for the child to scratch the itchy anal region, pick up the eggs under her fingernails, and sprinkle these invisible eggs in her path, on towels, on bedsheets, in the air.

They are swallowed by the child herself, other children, or adults and hatch in the small intestines, maturing into half-inch-long females and sixteenth-inch-long males. They mate and the males soon die. The fertile females, bloated with eggs, congregate in the bowel, eating bacteria and epithelial cells, waiting for nightfall to crawl from the anus and begin their itchy migrations.

Warning: People who read stories like this have a tendency to avoid shaking other people's hands. This can lead to loss of friends and a hermitlike life.

Since the eggs are not usually found in the feces, another method of diagnosis has been found. How can we sample the perianal folds to find the eggs? Simple. Turn the child over and press scotch tape on the area between anus and vulva. Place the tape under the microscope, and thousands of tiny, transparent eggs will be visible.

Despite the fantastic odds against reaching a host, more than three quarters of the world's citizens are or have been infected with roundworms using the methods of infection mentioned above.†

---

*They can also crawl into the female urogenital tract, get confused in this unprogrammed area, encyst, and cause trouble—even sterility.

†Crompton (1999) estimated that the world's human population was afflicted by 4.45 *billion* cases of infection with worms. October 12, 1999, is recognized as "six billion day," the day that the human population reached six billion. On that day Crompton's estimate yields an astonishing 74 percent human infection rate.[10]

PLATE 6

A. Sketch by Laurence Sterne (1885) from "A Sentimental Journey through France and Italy"; here signifies a child with multiple infections: pinworm (B, C), trichinosis (D), hookworm (E). He has survived almost universal dysentery of impoverished developing world. Will he survive his current infections?

B. PINWORM EGGS. Unusually small; dust-size; can be inhaled. Glassy double-walled shell. Egg at upper left undeveloped; others contain juveniles.

C. PINWORM, *Enterobius vermicularis*. Males 1/4–3/8 inch, females to 1/2 inch.
Female has long tapering tail; body transparent; egg-filled uterus usually contains about eleven thousand eggs; in rectum, crawls from anus at night, lays eggs in host's perianal area, then dies.

D. TRICHINA WORM, *Trichinella spiralis*. Males to 1/16 inch, females to 1/8 inch. Female produces live young that find muscle cells, convert them to nurse cells, and live *inside* them. Host muscle must be eaten for worm to become adult. Juveniles coiled up in capsules; longitudinal lines are muscle cells. Largest *intra*cellular parasite.

E. HEAD OF HOOKWORM, *Necator americanus, Ancylostoma duodenale*. Males to 3/8 inch, females to 1/2 inch; male has posterior fanlike clasping organ. Drawing shows strong teeth used to eat through wall of small intestine. Worms very active, leaving bleeding holes where they have been, exacerbating hookworm anemia.

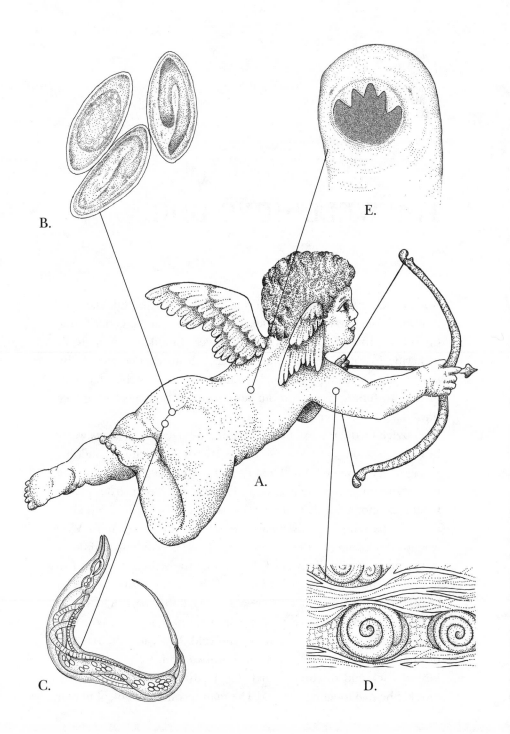

B.

E.

A.

C.

D.

7

THE WELL-HUNG DOG

A BLUE-GRAY PALL OF SMOKE and stench hung over the tuk-tuks, tiny motorized rickshas that are the taxis of the masses in Bangkok. Traffic lights acted like starting gates at horse races, the tuk-tuks frenetically jockeying for position. The wait for the light to change seemed interminable, as disproportionately loud roars and dense bluish clouds of exhaust fumes smothered our senses. A new smoke-belching race started at each corner.

Frazzled by the intense heat and noise, we decided to take a water bus upriver to our next destination, the Wat of the Reclining Buddha, a magnificent golden-domed temple at the river's edge. The "bus" is a boat about the same dimensions as a city bus, and it roars from station to station, momentarily touching a corner of its stern to the pier to allow the passengers to disembark. We reached our destination. Wise wife looked askance at the corner of the boat, bobbing up and down as the passengers disembarked. In the worst way she didn't want to jump from boat to land. We hesitated. The last disembarking person crossed over. New passengers were about to climb aboard. The sound of the engine changed to a mighty roar. It was now or never. Wise wife closed her eyes and stepped across the void. Unsteady, she reached out for support and grasped a saffron-robed arm. Safe on land, she looked back and discovered that the steadying arm belonged to a monk! She had touched a monk! The man recoiled. A look of horror distorted his face. The tourist guidebook had expressly discussed the celibacy of monks and the cultural prohibition against a female touching one. She was very upset. "What if I have ruined his career?" "What if I caused him to be a sinner?" she asked seriously. I said, "Buddha will forgive him." But she was not convinced. This was the time for a hug of consolation. As I administered the hug, I glanced around to see if I should be embarrassed, if we were breaking yet another taboo. No one was watching with disapproval at this Western-style demonstration of affection.

As I looked over her shoulder, I noticed a brown-and-white dog standing nearby, looking away at something. "My," I said to myself, "that is one well-hung dog." The dog's testicles were hanging down, literally, to the ground. I was sure its reproductive potential was diminished with every step. This idle speculation was interrupted by an epiphany: this animal has elephantiasis.*

THE INVADING ARMY

Elephantiasis, arguably the most visually spectacular of parasitic diseases, is caused by roundworms (nematodes). Covered by a plasticlike sheath, they are extraordinarily invulnerable to the defenses of the host. In most nematode infections, the parasite enters the host through the digestive tract, using resistant eggs to brave the churning stomach acid. But there are other avenues of entry. The skin, exquisitely designed to ward off environmental insult, is buttressed by backup blood-borne immune system armies—a virtual Maginot Line of defense bristling with weapons. Not discouraged, some nematodes have chosen to challenge these protective fortifications directly. Using the same strategy employed by Nazi armies in World War II, they

---

*Recently I learned that dogs do not get elephantiasis. No reservoir hosts for *Wuchereria* have been reported. The condition was probably testicular or scrotal cancer—or perhaps the dog was unusually virile.

Dogs do have a filarial nemesis, heartworm, common in the southern United States. It is caused by *Dirofilaria immitis*, a threadlike roundworm up to a foot long. Mosquitoes are the vectors. The adults are often found in the right ventricle of the heart, causing malfunction of the valves. Dogs sometimes die of heart failure.

bypass the imposing wall of skin. Miraculously, they have recruited allies to assist the penetration. All manner of terrible tiny terrors, primarily mosquitoes, are the guerillas in this invasion. The mosquito bites and releases live nematode juveniles on the skin. They burrow into the bite wound, thus bypassing the first line of fortification, the epidermis. The battle is joined.

Reserves are pulled up. Defensive armies of white blood cells engage. They use every manner of defense—poisonous antibodies, white blood cell infantry with gaping protoplasmic maws—to engulf the baby worms. But the worms fight back, defended by their invulnerable sheaths. Sometimes they cannot be detected by the host's immune system radar. In this ancient relationship, the parasite has evolved along with its host so that it is virtually immunologically identical to the host's tissues.*

THE GIANT SCROTUM

The adult nematode causing elephantiasis is four inches long and thread-thin. How can a quarter-inch-long mosquito introduce such a comparatively huge worm? Impossible. But the vector mosquitoes can exchange *microscopic juveniles* along with the blood they drink, in a lose-lose deal for the host.

Months (even years) after the introduction of the infective juveniles, the human host feels unwell and is routinely tested for malaria or the other diseases of the impoverished tropics such as pneumonia, tuberculosis, dengue, and AIDS. No malarial parasites inside the red blood cells. All tests are negative. The victim feels better. It was nothing. The books are closed. Then alternating chills and fever shake the victim and a new symptom appears, swollen lymph nodes. *It must be malaria.* A thin film of blood is stained on a slide. Still no malaria parasites in the red blood cells. But wait! In between, but not inside, the

---

*The idea of parallel evolution has been buttressed by new DNA evidence demonstrating that mutualistic bacteria living in the gut of stink bugs have evolved along with the bugs. At each evolutionary step producing a new bug species, there was a comparable evolution of a species of gut bacteria. This information demonstrates that parasites can evolve along with their hosts.

blood cells is a horde of wormy, squirmy swimmers, the offspring of the threadlike females. The diagnosis is clear—filariasis, that dreaded distorting disease. Malarial parasites survive by hiding in red blood cells, only momentarily vulnerable to antibodies when they float free to infect new cells. But the baby filarial worms fight it out in the open. They live in the plasma of the blood, not the cells.

Adults live in lymphatic spaces, the thin-walled chambers and threadlike processes under the skin and throughout the inner body. They converge into pea-size structures—lymph nodes. The vessels and nodes run the length of the digestive tract and accumulate in the groin and armpits. The fluid in the nodes is called lymph and is identical to the watery plasma of the blood. The plasma leaks from the circulatory system into the parallel lymphatic system. Lymph often accumulates in damaged parts of the body, becoming the "water" inside blisters and causing swelling around injuries.

The lymphatic system culminates in a large, thin-walled vessel, the thoracic duct. When lymph flows into the heart through this duct, it becomes plasma again. But there is no force to push the lymph back to the heart. What motivates its flow? Muscle power. When leg muscles contract, they shorten and become rounded. The momentary bulge squeezes the lymphatic vessels and forces the lymph up toward the heart. Tiny valves prevent the lymph from flowing backward. It is this slow flow that makes us vulnerable, providing a tranquil backwater for the adult filarial worm.

Elephantiasis-causing *Wuchereria bancrofti* is one of the most spectacular species of human-infecting filarial worms (another is the musical *Loa loa* mentioned in the introduction). *Wuchereria* does not chew away at one's innards or suck up nutrients. Instead, it harms by using the body's own immune response, inflammation, to cause swelling and lay down connective tissue.

The worms can live for twenty years. In a few long-term cases, spaghetti tangles of adult worms inflame the tissues around the lymph vessels, causing them to swell permanently. The surrounding tissue is stretched and stressed. The body's defenses respond with hyperplasia, closing off the threat with an overwhelming avalanche of cells, and hypertrophy, an unnatural cellular enlargement, creating new tissues that pile up to form a grotesquely swollen leg and scrotum. If not cared for, leg and scrotum become huge—elephant size.

The awful bulge expands the dimensions of the leg and obscures the toes, exposing only the toenails, like the leg and foot of an elephant—hence the name. The lymph nodes in the groin may expand likewise until, over twenty or so years of untreated infection, the unfortunate victim has to wheel his scrotum around in a wheelbarrow. This is not a sexually specific disease. There are classical photographs of a woman with a single monstrously enlarged breast.

THE MYSTERY

Females give birth to living young. The tiny, wormy infants are huge compared with red blood cells. They float along in the blood, eventually reaching the heart. Each ventricular pulsation pumps them into the universal parasite nursery, the oxygen-rich lungs. They grow feverishly each day, resting during the night. But while they rest, they relax their hold on the vessels of the lung and are washed out into the blood in uncountable numbers.

A blood sample is taken between 10 p.m. and 2 a.m. The stained slide reveals microfilariae interspersed among the blood cells. The diagnosis is made. The patient is given an anthelmintic and goes home, cured.

But a mystery popped up. After World War II many indigenous people migrated from remote places in the interior of Pacific islands like Borneo to get jobs. Among them were men and women with grotesquely swollen legs, breasts, and testicles. Modern medicine had not penetrated into the mountains; these were dramatic cases of advanced elephantiasis. Some infections were twenty or thirty years old. Technicians took blood samples at the traditional times, between 10 p.m. and 2 a.m. No microfilariae. But these people were obviously infected. Why was it seemingly impossible to make timely diagnoses of elephantiasis on these Pacific islands when it was so easy on the mainland? The mystery defied solution.

The answer came accidentally, as often is the case. A public health parasitologist, working during the day, took a sample of blood and stained it. It contained microfilariae. Why weren't they apparent before this? Answer: no one looked for microfilariae in the daytime. They were supposed to be present at night. But there they were on the stained slides. It became apparent that *the cycle was reversed on the Pacific islands*. The baby worms, hidden in the lungs at night, floated free during the day. That was the solution to the mystery.

But why, in Fiji, Samoa, the Philippines, and Borneo, were the microfilariae present primarily during the day and not at night like any self-respecting *Wuchereria* infants? Research revealed that the common mosquito on the Pacific islands, *Culex pipiens*, takes its blood meal during the day. If the microfilariae were in the blood at night, the mosquitoes would not have access to them. They would be unable to continue their life cycle. The parasite had evolved to match the behavior of its daylight-feeding vector in order for its life cycle to be complete.

Evolution is neither directional nor purposeful. It is likely that the microfilariae were originally present in the blood day *and* night. In areas where the mosquito was nocturnal (the mainland), only those that were in the blood at night were sucked in. They were then transmitted to human hosts where they matured and reproduced. The genes for daytime presence in the blood were eliminated. Hence the nighttime cycle. The mosquitoes on the islands were daytime feeders, hence the selection for diurnal microfilariae in the blood.

By the way, how do the microfilariae in the lung know when it is nighttime?*

As for the well-hung dog—imagine a spectacular neutering scenario.

CAN THERE BE A PARASITIC DISEASE MORE GROTESQUE THAN ELEPHANTIASIS?

It was a classical African mind-numbing scene. Appearing like black ghosts in the early morning haze, a line of tall, gangly, emaciated elderly men formed a miserable parade. In the early morning light

---

*It has been shown that when the host's sleep periods are reversed, the presence of microfilariae is reversed. Experiments demonstrate that in the presence of high oxygen concentrations (during sleep), the larvae accumulate in the lungs.

their stick-figure silhouettes were all bony knees and elbows. Hand-in-hand they staggered. They were being led by a child to a cool place under a tree near the banks of a river to spend the day in unseeing repose. They were blind.

It was the river that was their undoing. They were suffering from a disease called river blindness. The blackfly, *Simulium*, rears its young in the river. Only swiftly flowing rivers are suitable for the development of the juvenile flies. Upon hatching, the fully developed infant crawls out of the water and flies off, seeking the blood meal that will sustain it and provide the nourishment to produce eggs to continue the cycle.

Blackflies have not evolved the tubular sucking mouthparts of their better-adapted brethren. They have to stretch the skin and make a tiny cut to create a pool of dissolved skin cells, lymph, and blood that they suck up along with microfilariae of the nematode *Onchocerca volvulus*. After undergoing development in the fly and becoming infective juveniles, they are introduced into another human host by the fly's saliva when it feeds again. The juveniles wander under the skin and gather, forming nodules where they mature and mate. Eggs hatch inside the female and she releases newborn microfilariae one by one. Each female can produce a thousand a day, and many females produce a horde, invading subcutaneous tissues in writhing, wandering masses.

Here is where the cycle of *Onchocerca* differs from *Wuchereria*. Often the adults are overtaken by the host's immune system and are interred in a calcified or fleshy nodule. In Africa, these nodules project prominently below or along the waist, ominously signifying the possibility of eventual blindness.*

Surviving adult males are much smaller than the females, which can reach *two feet* in length. Entangled mated couples can often be found in the fleshy tumors. Females release microfilariae that migrate in hordes under the skin. In laboratory studies it has been found that these microfilariae are attracted to light, although they have no eyes.

---

*In Central America there is a tendency for the nodules to be located above the waist, especially on the cheekbones and forehead. It may be that there is a greater likelihood that the microfilariae will invade the eyes due to their greater proximity (Don Duszynski, 2008, personal communication).

It is not this ability to respond to light that causes them to migrate through the eye in their immense numbers. It is simply the volume of the migrating masses under the skin. Thousands end up in the eye. Their paths are followed by white blood cells, attackers that mobilize the formidable immune system of the host. The living microfilariae do not elicit a strong reaction, but once they die, they rot. The body's white blood cell vultures are attracted to the carrion. The cornea of the eye is often the scene of a massive feeding frenzy. White blood cells and antibodies mop up fragments of dead microfilariae. The typical progression of this inflammation ends in the production of scar tissue. Tiny calcified scars cover the cornea, remnants of these immune reactions. The retina and optic nerve are sometimes involved. Each female worm can pump out young for up to sixteen years, and the microfilariae keep wandering into the eye.

A telltale whitening of the cornea in infected teenagers is a harbinger of eventual blindness. Twenty or thirty years and repeated infections bring about total blindness. By age forty or so, the host's cornea has become a densely white, opaque worm cemetery. It takes that long for the immune system to achieve the inevitable horrible consequence.

The dead larvae under the skin also wreak havoc. Their death is accompanied by a violently itching rash and often a loss of skin elasticity, producing characteristic wrinkles and a ghastly appearance of premature aging.*

The slave trade in the sixteenth century brought a uniquely African form of revenge to the Americas. The blood of infected African slaves was just as delicious to the local blackflies as African species, and the life cycle of these filarial worms was introduced to the New World. In some areas of Africa, 90 percent of the population contain microfilariae in their blood, so a huge proportion of the slaves must have been carriers. Today, of the forty million cases in the world, almost a

---

*Other symptoms, including enlarged scrotal region, hernias, and cracking skin, are also produced by *O. volvulus*.

PLATE 7

A. WELL-HUNG DOG. View seen while hugging wife in Bangkok.

B. WELL-HUNG HUMAN. Dog's pseudo-virility cannot be attributed to *Wuchereria bancrofti,* but man's can; he probably had elephantiasis for twenty years. Adult worms in lymph spaces in groin clogged lymph vessels; cellular reaction caused permanent enlargement.

C. ELEPHANTIASIS CAUSED BY *Wuchereria bancrofti.* Males to 2 inches; females to 4 inches. After mating in lymph spaces, females release sheathed juveniles, microfilariae, that migrate through body's lymphatics, eventually entering blood. Microfilariae ingested by mosquitoes infect next victim. Man reported he was unable to walk; he was shunned by villagers.

D. MICROFILARIA of *Wuchereria bancrofti.* Flexible sheath not visible; dots are stained nuclei; juvenile about twenty-five times the diameter of white blood cell.

E. RIVER BLINDNESS. Classical picture of blind elderly men being led to river's edge by boy, to sit all day. Caused by filarial worm *Onchocerca volvulus.* Adults entwine under skin and are encapsulated by host, forming prominent nodules that calcify after prolonged infection; eyes damaged by long-term allergic reaction to microfilariae.

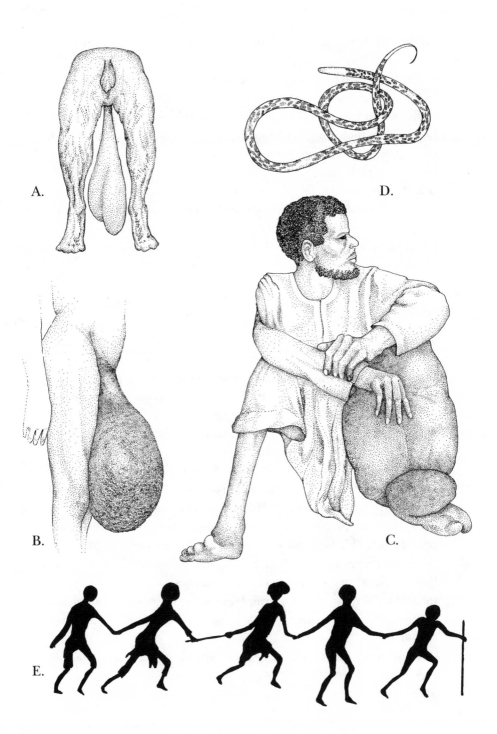

million are found in Central and South America and other tropical areas with a history of slavery.

As of 1975 two thousand cases of early onchocercariasis were found in London (in immigrants, presumably).

In the northeastern United States, trout fishermen are besieged by blackflies in the spring. Uncountable hordes attack the unprotected, driving them away from the streams, swollen and itchy. Experienced fishermen swath themselves in netting, but although they wear gloves, bracelets of red welts on their wrists where glove meets sleeve are testimony to the ability of the flies to defy even the most careful preparation. The vectors are here. But we Americans are protected by the temperate climate, inhospitable to these tropical parasites.

But consider global warming. . . .

HOPE

There has been a massive international effort in recent years by pharmaceutical companies to assume a major responsibility to combat tropical diseases in impoverished countries by donating drugs at little or no cost. For example, the 2007 report of the British company Glaxo Smith Kline contains this passage: "January 2008 saw the 10th anniversary of our commitment to eliminate lymphatic filariasis (LF), also known as elephantiasis. To date we have reached over 130 million people, and 24 million children have been born in areas that are now LF free. . . . The lymphatic filariasis elimination programme continued . . . making almost 750 million treatments in total."

Similarly, in the 2007 report to shareholders of the American company Merck: "It's been 20 years since Merck launched its Mectizan Donation Program. Since then, Merck has donated more than 1.8 billion Mectizan tablets to patients in 33 countries, helping protect more than 69 million people from river blindness every year. . . . Merck has reaffirmed its pledge to donate enough *Mectizan* to eliminate river blindness."

Virtually all major pharmaceutical firms offer programs such as those above.

8

FIERY SERPENT

AN ALMOST IMPERCEPTIBLE BREEZE created sky-colored rippling reflections serrating the serene surface of the papyrus-edged pond. The
early morning sun reflected from the huge river in the distance—a
silvery slice slashing the savage desert with a knife of green. A few dun-
colored, squarish houses formed a crescent around this pool spawned
by an eddy of the river, their pale mud bricks speckled with fibrous
straw. Water-filled irrigation ditches, scarcely wider than a few paces,
radiated from the shores of the pond.

A girl was washing white galabiyas, enveloping robes designed to
fend off prying eyes with their voluminous folds and simultaneously
provide whatever flow of air possible in the already stifling heat. The
water felt cool. She watched as an annoying blister on her arm became submerged, hoping that the cool water would alleviate the discomfort it caused. Suddenly, the blister burst! Before her horrified
eyes a threadlike, undulating loop emerged from the wound. From
a tear in the loop's surface erupted a whitish fluid that clouded the
water. The girl cried out more from shock than pain, staring at the
still undulating, sinuous loop with fear. An old lady shrouded in an
impregnable-looking, robelike black abaya emerged from the dark
shade under a palm tree like a walking shadow. At her beckon, the
crying child approached, looking for solace. Wordlessly she took a
thin splinter of palm leaf and placed it on the child's arm. Carefully

grasping the spaghetti-thin thread in her spindly, shaking fingers, she pulled. After a few light tugs a featureless, pointed tip appeared. The worm's head waved back and forth blindly, as if it was shocked by the bright outer world and wanted to return to its dark haven under the girl's skin. It began to retract. But the old woman had fought this battle before. Murmuring softly to reassure the child, she wrapped the head and first inch around the twig. She released the child's arm, the suspended worm dangling. Still crooning, the grandmother enshrouded the girl in the voluminous folds of her garment and whispered ageless advice: "Every day tug at the worm and wrap an inch around the twig. But be very careful not to pull too hard or it will break and that will make you very sick." She knew that if the worm was torn it would die, creating a festering wound the length of the child's arm.

A few weeks later the family was bringing their date harvest to town. The little girl wouldn't go. The worm-wrapped stick, now a mound of entwined once-living spaghetti, hung from her forearm. She had not finished winding, although a yard of worm dangled from her arm. She was too embarrassed to be seen.

THE FIERY SERPENT

And the Lord sent fiery serpents among the people. . . . Therefore the people came to Moses and said, we have sinned . . . pray unto the Lord that he take away the serpents from us . . . and Moses made a serpent . . . and put it on a pole.

—NUMBERS 21:6–9

The fiery serpent afflicted man for millennia before the Bible. Progenitors of modern humans evolved more than two million years ago. It must have taken much of that time for the worm to find its devious way to enter the human body. The complex life cycle of the Guinea worm, *Dracunculus medinensis*, reveals how humans inadvertently cooperate with the parasite, drawn into an unwilling relationship without knowing it. But there is another unwitting participant in the circle

of infection, an almost microscopic mini-shrimplike animal, the ubiquitous one-eyed copepod, *Cyclops.**

The little girl's blister begins the cycle. The unbelievably long, spa-
ghetti-thick female worm has mated somewhere in the darkness of
the child's abdominal cavity connective tissues. The worm migrates
to the undersurface of the skin, and thousands of eggs hatch inside
her three-foot-long body, filling it almost to bursting—a feeling, I
imagine, not unlike the ninth month of human pregnancy. At some
physiological signal initiated by cool water, she moves up from her
under-the-skin lymphatic lair and produces an irritating fluid that
causes a blister on the skin.[†] The blister bursts. In great pain, the child
submerges her arm. The alien mother continues her unthinking response to the watery stimulus. A loop of her larva-laden uterus bursts
from the center of the suppurating lesion and explodes, releasing half
a million writhing microscopic juveniles into the water. For months or
years more, she continues to pump out thousands of juveniles.

Dozens of swimming dots, *Cyclops*, fix their one unseeing eye on the
source of light, the water's surface. Inadvertently they enter the cloud
of microscopic squirming slivers that are the baby worms and filter
them from the water, carrying them in a ciliated groove to the mouth.
The infant worm is swallowed, its protective sheath preventing digestion in the gut of the copepod. Inside, it "knows" just what to do, eons
of evolution providing a genetic map of the innards of the copepod.
Unthinking, the juvenile follows its genetic directive and migrates into
the body cavity of the copepod, where it molts three times—and waits.

When the little girl dipped the drinking-water container into the
pond, she hadn't noticed tiny, erratically swimming motes of life, the

---

*Copepods are only remotely related to shrimp. They are herbivorous minute crus-
taceans that feed by filtering tiny algae from the water with their hairy antennae. Secondary vortices produced by legs allow the worm larvae to enter the mouth. Copepods
may be the most common animals on Earth, surpassing the unbelievably ubiquitous
nematodes.

†Some data suggest that the blister is caused by an allergic reaction to waste products or larvae released under the skin by the female.

PLATE 8

A. Drawing by Sandy Chichester Rivkin (2008) depicts Egyptian girl washing clothes in pond near Nile. Jug contains water scooped from same pond. A blister on her submerged arm tore open. Section of worm erupted from lesion; its uterus burst, freeing a white cloud of microscopic juvenile worms that were eaten by copepods. Girl drank copepod-contaminated water from jug and became infected.

B. GUINEA WORM, *Dracunculus medinensis*. One of largest nematodes; females to 32 inches, males very small, to 1½ inches. Female's uterus fills with juveniles, creating great pressure; gut squashed and nonfunctional. Worm extends up arm under skin; its head (or larvae) releases irritant, forming blister that bursts in water. Distended uterus smashes through worm's body wall and bursts, freeing juveniles.

C. LESION. Worm was removed; healed wound pinkish with reddish ridges; up to 2 inches across. In Sudanese village 20 percent of population had lesions; 28 percent of these had multiple lesions (1996).

D. TRADITIONAL METHOD OF REMOVING WORM. Each day, stick is gently turned. Takes weeks or months; worm dries during process.

E. WORM PROTRUDING FROM HOST'S SKIN. Uterus bursts, releasing up to half a million juveniles; copepods eat them. Microscopic juveniles, visible (enlarged) in water above copepods, migrate through gut into copepod's body, molt three times, become infective when copepod is ingested by human.

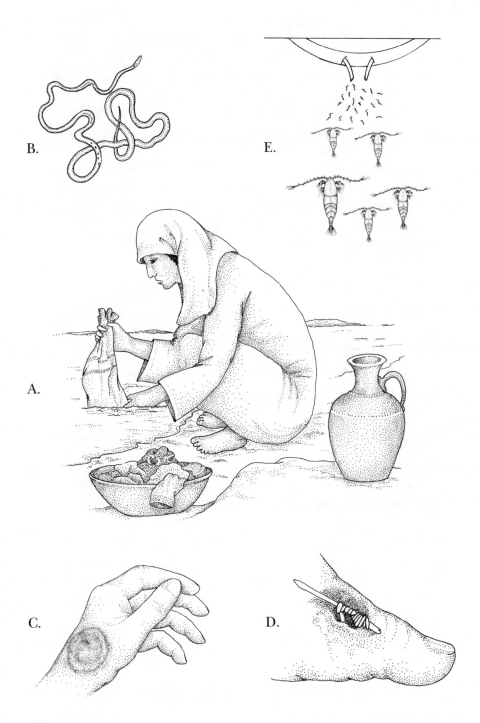

B.

E.

A.

C.

D.

copepods. She drank the water with its malevolent living contaminants. Once they reached the first inches of her small intestine, the copepods were digested and the ensheathed guinea worm juveniles safely freed. About twenty days after being swallowed, the juveniles lost their sheath and began a programmed migration, burrowing through the intestine wall. After two molts they ended up in a capsule in the liver, in the abdominal cavity, or under the skin.

After a while the now-mature female leaves her haven and mates. Eight months or more elapse. Her uterus swells with eggs that hatch into a squirming mass of thousands of young. She feels the urge and migrates toward the skin surface and gives birth to her writhing burden.

AMERICAN STORIES REQUIRE HAPPY ENDINGS

The little girl faces death. The hole the mother worm produces remains open even after she is removed. It abscesses, causing such intense pain the little girl cannot do her chores. Finally she is taken to town where modern medicine cures her. She is comparatively lucky. On occasion the worms have been known to become confused and migrate throughout the body to die and become covered with a chalky tomb produced by the victim's body as a defense mechanism. Rarely, wandering worms have been found encysted in the spinal cord, causing paralysis.

Sadly, our heroine has no immunity to the worm and can become infected again and again. Will she survive or not? Tune in tomorrow.

9

IT HARDLY EVER HAPPENS

I HAVE ALWAYS WANTED TO EAT SEA URCHIN EGGS—don't ask me why. Perhaps for their exotic novelty; perhaps for their famed aphrodisiac properties; or perhaps because I expected that they would make me rich. The opportunity came when I found out that a commonly eaten sushi dish called uni is composed of raw sea urchin roe wrapped in seaweed. For a non-sushi-eating kind of guy, the dish is daunting, but I managed to swallow the rosy-gold, egg-laden caviar and its wrapping. "Not bad," I thought—slightly fishy with a crisply oceanic aura and a good nose. I cannot report here on its efficiency as an aphrodisiac, but it was a disappointment on another level. It didn't make me rich. The roe, valued at over a thousand dollars a pound in Japan, is literally fishy gold. But its golden pink color was my undoing.

The bottom of the turtle grass community in the Caribbean is strewn with the "sea egg urchin," *Tripneustes ventricosus*. Black, covered with small white spines, they glimmer like three-inch fuzzy hemispheres as, in their profusion, they punctuate the green, grassy bottom. So abundant is this animal that in some cases one can gather thirty or forty in ten minutes. In season the urchin is filled with thousands of eggs that on most Caribbean islands are considered eminently powerful in their capacity to increase male potency. (In the Caribbean almost everything eaten is said to increase male potency.) Sadly, although they are easily harvested and have the required potency, the eggs are greasy black in color. The roe popular in Japan is extracted from the flamboyantly named green sea urchin, *Strongylocentrotus drohbachiensis*.* It is distinctly rosy. The picky Japanese market will not accept plain black roe. My dream of becoming megarich was not to be. I was destined to go back to my slightly frayed life as a professor.

My culinary foray had one encouraging result—the taste of sushi became appealing. But a long-buried memory kept mysteriously surfacing when I thought of eating raw fish. "Raw invertebrates yes, raw fish, no" was my dim recollection. Then the memory popped to the surface of my consciousness—raw fish—anisakiasis.

OOSTDAM

That's his real name. He is Dutch. One day in Amsterdam, Oostdam was vigorously extolling the virtues of his hometown. "One delicacy must be eaten to complete your experience. Follow me." He rushed off down the cobblestone street until he found a vendor with a cart. Dead fish festooned the hooked metal frame surrounding the ice-filled cart. Their odor advertised the vendor's wares. The delicacy was lightly salted raw herring, evidently the national dish. "We came three thousand miles to walk down a smelly street to eat raw fish?" asked a skeptical friend. Tentatively, I tasted the slippery gray-tan flesh. The same inchoate memory surfaced—anisakiasis—and I handed back the dead fish, to Oostdam's dismay. One cultural experience would remain incomplete.

SUSHI, SASHIMI, CEVICHE, OR SALTED HERRING— TAKE YOUR CHOICE

Accidents happen. In this case, a seal defecates—not an accident, you say?

But the accident will occur long after and far away. The seal is infected with a nematode, *Anisakis simplex*, and a chain of events is initiated that will end up with a gut full of worms—not of another seal,

---

*Possibly the longest name in the taxonomist lexicon.

but of an ardent sashimi or raw herring eater. How can the feces of a seal in Arctic waters cause a horrendous parasitic condition in a raw fish eater in Tokyo, New York, or Amsterdam? In this case a little feces goes a long way. An infected fish is caught by a fisherman in the far North and taken back to his home port.

It happens this way: the seal's feces contain hundreds of tiny glassy-shelled eggs. They are ingested by various microscopic filter-feeders like our ubiquitous copepods. These are eaten by small fishes, which are eaten by bigger fishes, and so on. By the time the next seal eats the biggest fish in the food chain, the number of worms in the fish's body is magnified many fold. The seal gets a mouthful of juvenile worms in its fishy meal.

If you interfere with the normal seal-fish-seal cycle, you will become accidentally infected—a proxy seal, since the larval worm buried in the flesh of the herring cannot differentiate between your gut and that of a seal or killer whale.

BUTTING IN

A huge net is dragged through the icy waters. Its contents are spilled on the deck in a silvery harvest. If the fishes are tuna, this evokes frenzied activity as the crew attacks each fish with sharp knives. They must be bled immediately. A giant tuna can be worth fifty thousand dollars, and care must be taken to preserve its exquisite taste. Later it is eviscerated. Sometimes the crew is not quick enough. Thousands of juvenile worms have wandered from the cold intestines into the tuna's warmer muscles.

By contrast, the less-valuable herrings are shoveled into the fish hold and covered with ice, and not eviscerated at all until they reach the packing plant. Occasionally a herring's body is riddled with chubby juvenile worms. The fish is just the middleman, passing on the larval worms to the mammalian definitive host—normally a fish-eating mammal—a seal, porpoise, or whale.

Icing makes the worms cool. It doesn't kill them. In fact, for some reason, cooling causes these juvenile worms to migrate from capsules similar to cysts located in the abdominal cavity into the fishy host's tissues.

PLATE 9

A. CALIFORNIA SEA LION, *Zalophus californianus.* Definitive host of *Anisakis* spp. Seals, fish-eating whales also hosts; acquire worms from eating infected fishes.

B. *Anisakis* EGGS, released in sea lion feces, hatch; juveniles float freely in water; eaten by swimming shrimplike crustaceans.

C. COPEPOD is symbolic; many small crustaceans can serve as first intermediate host, e.g., krill, amphipods.

D. ATLANTIC HERRING, *Clupea harengus.* To 18 inches; one of most common fish in world; eats copepod, becomes second intermediate host; many alternative fish hosts, e.g., cod, salmon, flounder, bluefin tuna. Juvenile burrows through fish's gut wall, coils up and encysts in muscles or on outside of intestine, waiting to be eaten by whale or seal.

E. HUMAN EATS INFECTED FISH and (rarely) gets anisakiasis. Most infections are not noticed, and worms pass out of body because they cannot survive in humans. Sometimes juveniles, freed from cysts when fish tissue is digested, wander through human body, penetrating stomach or intestinal wall. (E) shows three (inch-long) juveniles projecting into intestinal lumen as seen through endoscope, which was then used to remove them.

F. LARGE X symbolizes that human infection is accidental—a dead-end for the worm because it requires *marine* mammal to complete its life cycle.

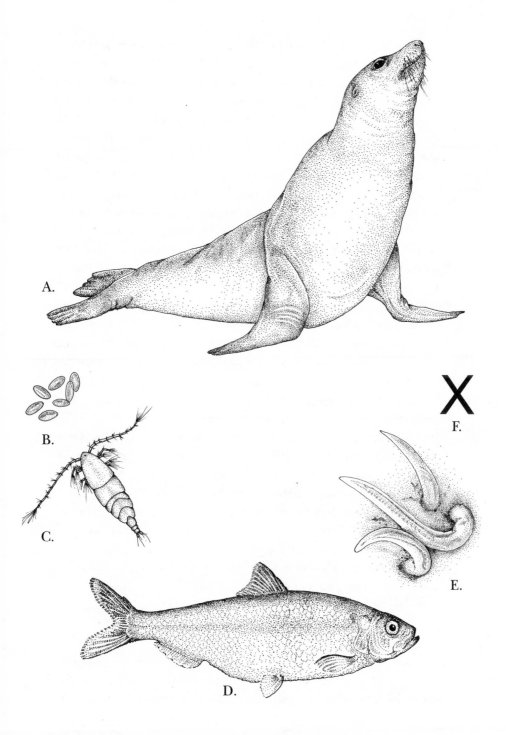

A.

B.

C.

D.

E.

X

F.

Unlucky Japanese sushi or Dutch herring eaters are accidental victims.

Humans are not supposed to be involved in a fish-to-seal or fish-to-porpoise parasite life cycle. Butting in exacts a price. The intruding human victim can experience appendicitis-like pains as the juvenile worms migrate to the intestine or stomach wall seeking the richness of seal or porpoise gut—only to be frustrated by this unnatural human intestine. Sometimes the symptoms can be mild or the infection can be asymptomatic as the human, not sensitized to these new invaders, has not had time to build up an immune response. More frequently the juveniles burrow into the lining of the human host's intestine and provoke the body's defenses. The invader is soon surrounded by a spherical tomb of granule-filled large cells (a granuloma), or mucus. The juvenile writhes around in the mucus for a while but dies quickly in the granulomatous lesion. Eventually the center of the granuloma disintegrates and pieces of dead worm float free.

In these cases the body's immune system has attacked the embedded worm. White blood cells create the lesions and the symptoms. On rare occasions the undetected, unhindered juvenile kills the human host as it burrows through the intestinal wall to cause peritonitis. Raw-fish eaters sometimes see fat, wiggling juveniles in the fish flesh they are eating and are so disgusted that they throw their lunch away. These are the lucky ones. Others, who eat less-obviously infected fishes, become incidental carriers (paratenic hosts) in lieu of the fishes.

The juveniles cannot mature to adulthood in the human body.

It is reassuring that *Anisakis* or its close relatives, *Contracoecum* and *Pseudoterranova*, normally reach sexual maturity in porpoises and seals, not humans. We are simply enigmatic mazes to lost juveniles that cannot find their way, destined to undulate futilely in a bloody pit in the human intestinal lining until they die.

FRUSTRATED PHYSICIANS

There is an odd usage of the word "lucky." If someone is infected, he or she is "lucky" to be afflicted with the most obvious and ghastly clue to the disease—vomiting up a fat, inch-plus-long worm. Even the most naïve physician can diagnose this unfamiliar condition. But absent this disgusting symptom, gastroenterologists in the United States are often frustrated. They have been called upon to diagnose a putative case of appendicitis. They end up needing a TV-style genius because X-rays do not show a swollen appendix. Hopefully the real-life medical hero calls for a CT scan and recognizes confusing shadows of elongated, thick strings.

But if the genius is not available, the last resort is to view the inside of the intestines through the flexible, flashlight-tipped endoscope. At last, the cause: undulating, spaghetti-like worms projecting into the lumen of the gut. How to remove them? If no clear-cut method is available, operate. In a typical understatement, a physician/researcher reported, "Removal of the worms with biopsy forceps is effective, although catching the lively worms with a forceps may be challenging."[11]

Medication is much preferred.

Japanese sushi and sashimi eaters are most vulnerable by virtue of their frequent ingestion of raw fish; the Dutch and Scandinavians are less commonly infected because recently enacted laws require herrings to be deep-frozen. But Americans come a close fourth, immigration providing exotic foods and exotic parasites. It has been reported that more than 80 percent of raw fish on the U.S. west coast contained *Anisakis* or its relatives.[12] On the east coast *Pseudoterranova* is called the "codworm."

On five occasions fresh salmon were purchased from supermarkets in Michigan. All contained salmon with encysted juveniles.[13]

But anisakiasis isn't common; only about a thousand cases were reported last year in Japan.

That makes the odds 1:128,000. Not bad, sushi lovers.

10

THE ANTI-SEMITIC
TAPEWORM

How awful is that ancient form of racism, anti-Semitism. Is it possible to go to further lengths of evil? Horrifically, yes. In the animal kingdom a monster adds sexism, a dollop of malevolent anti-female bias. Then it adds to this hellish mixture a particularly terrible ingredient—it attacks the elderly. Anti-Semitism and sexism, not only against Jewish women in general, *but against elderly Jewish women.*

Nature abhors no form of behavior. Morality, as created by humans to curb their behavioral excesses, cannot be imposed on other animals. Yet the mind cannot but attribute the term "monster" to a parasite that singles out little old Jewish women as its prey. What explanation have we for this seemingly vicious interaction between the monster and an elderly woman? Why does it attack the innocent? How does it accomplish its abhorrent deed?

The answer begins in Europe, with the late-nineteenth-century massive migration of impoverished Scandinavians to the United States. They brought with them a parasite so terrible as to rival the legendary lamprey that ravaged the trout population of the Great Lakes. In Minnesota and Wisconsin, Swedes, Danes, and Norwegians found a familiar countryside and settled. The environs became familiar to another immigrant—the huge, broad fish tapeworm, *Diphyllobothrium latum*. Its complex life cycle requires the presence of minnows and carnivorous fishes similar to those found in Scandinavia. These it found in abundance, and another pioneer established itself in the bosom of America.

PASSOVER AND THE MODERN DAY ANGEL OF DEATH

Jewish festivals in America are presided over by the grandmother, often, in the past, a wizened old lady who spoke broken English. The menu invariably consisted of an appetizer, a ball of finely chopped gefilte fish (gefilte = filled), followed by chicken soup, followed by the pièce de résistance, boiled chicken, once removed (removed from the soup pot and then roasted).*

But this seemingly benign meal requires an explanation—no, an exposé. Grandma was so old that she had lost many of her senses, notably the sense of taste. So she followed the ancient traditions blindly— no, tastelessly. To extract the essence of chicken, which makes chicken soup the icon of healthfulness, she boiled the poor bird interminably. The soup, with pearls of fat floating on the surface, was the precursor to the main course: that same chicken, a poor, ravaged, pale shadow of its former clucking self. It was like eating a mummy (nowadays, in more prosperous times, modern moms use a separate chicken for the soup and roast another for the main course).

The assembled family would sacrifice their taste buds on the altar of tradition. But somewhere, sometime, someone discovered the antidote to eating this chicken library paste for holiday dinner. That hero discovered a boon to mankind—horseradish! The modern technique is to slather fiery horseradish on the first course, the gefilte fish, so as to destroy the taste buds, making the rest of the meal palatable. Real men choose the undiluted white version; kids, the red version that is mixed with beets. (Eating white horseradish is the real test of manhood, not the bar mitzvah.) Grandmother smiled benignly over the

---

*Wise wife tells me this is an exaggeration brought about by unpleasant recollections. Even in the old days a different chicken was roasted. It seems, in the taxonomy of cooking, that there is a kind of chicken known as a "roaster" and a kind known as a "boiler."

scene, noting tears coming from the eyes of the assembled—tears of joy, she thought.

In the process of preparing the gefilte fish, Grandma would season to taste, necessitating sampling a lot of the ground fish (because of her taste problem). This was the moment of disaster! Buried in the flesh of the fish was a larval tapeworm. But the worm parasitized only the rich, the former European peasants who, now in America, were able to afford a pure white version of gefilte fish. Why the rich? The poorest served a brownish ball made exclusively of that coarse fish, the carp. They did not become infected. The well-to-do mixed in white-fish and yellow pike. It was the yellow pike, *Stizostedion tritreum,* that the host-specific tapeworm utilized as its second intermediate host.

THE DEFECATING SCANDINAVIAN

Long ago, shrouded in the clouds of history, a Scandinavian immigrant to America defecated on the ground. Rain washed the feces into a lake. They contained the eggs of another settler from Europe.

The eggs hatched into a ciliated ball of cells resembling a microscopic hairy basketball. These larvae were strained out of the water by tiny, ubiquitous filter-feeding copepods, specifically of the genus *Diaptomus*—any old copepod would not do. The larvae hatched and penetrated the gut wall, taking up residence in the body cavity of the copepod. There their shape changed and they became infective.

The copepods were eaten by minnows, carrying forward the sequence of events in the complex life cycle. The copepods became the first intermediate hosts; the minnows, the second. An infected minnow, containing the waiting second larvae in its tissues, was then eaten by a carnivorous fish, the aforementioned yellow pike. The larvae did not develop further but grew into a featureless white, macaroni-like worm in muscles under the skin. There you have it, the source of the apparent prejudice of the worm! The parasite was transferred to the little old woman as she tasted the raw chopped yellow pike mixed with the carp.

The gefilte fish or broad tapeworm is an aquatic variant of human-infecting tapeworms. All tapeworms must have originated as water dwellers. In their quest for an unchallenged biological niche, tapeworms forsook the open waters for animal intestines. But their watery origins have not been lost, and larval forms are aquatic. More recent evolution produced clever tapeworms that have evolved the capacity to infect humans via terrestrial intermediate hosts. In a consummate stroke of pragmatic evolution, the pork tapeworm, *Taenia solium*, can infect people without an intermediate host. But the most common human tapeworm, the beef tapeworm, *T. saginata*, needs a cow. The cow is an intermediate host and the human is the definitive host in which the worm becomes sexually mature. It grows to up to fifteen feet in length. Humans do not normally get infected by eating beef tapeworm eggs. Cow intermediate hosts are needed.

Like the gefilte fish tapeworm, the beef tapeworm seems to be prejudiced. It "chooses" Turkish men in the same manner as the gefilte fish tapeworm chooses Jewish women. Few Turks are infected by the pork tapeworm. Why? Turks are Muslims. Only sinners eat pork. Why men? Rural Turks, herdsmen, have a penchant for eating rare beef chunks skewered on their daggers—sheesh kebab. In India, Muslims become infected with beef tapeworms while Hindus, who venerate cows, do not.

TAKING THE MEASURE OF TAPEWORMS—
TERRESTRIAL TAPEWORMS

The tapeworm body consists of a string of repeated, flat, segmentlike structures called proglottids, each apparently more or less identical to the other (internally they vary with age). There can be hundreds of proglottids in the string. In terrestrial tapeworms, each ends up as a bag of eggs that crawls or drops out of the anus as an individual proglottid or attached to others in a string of four or five.

Once on the ground, the proglottid rots, but its cargo of weather-resistant eggs remains to be picked up by an intermediate host, perhaps an insect like a flea or beetle. The intermediate host, in turn, is eaten by the definitive host (sheep, dog, man—most vertebrates). That is how terrestrial tapeworms avoid the ancient requirement of water for reproduction. Aquatic species like the gefilte fish tapeworm use the ancient technique and release the eggs through a birth pore, melding them with the host's feces that must eventually be washed into a lake or stream. The birth pore obviates the necessity of shedding the egg-filled proglottids, so this tapeworm stays intact and becomes huge. It can grow to thirty feet or longer, with inch-wide proglottids (hence its alternate name, broad fish tapeworm).

It is ironic that little old women harbor the largest of human tapeworms. It comes to mind, naturally, that the tapeworm causes the "little" old woman to be little. It has been proven, time and again, that most tapeworms do not extract an inordinate amount of food. It is rare that a tapeworm has an apparent effect on the host. While it doesn't remove enough nutrients to deprive the woman and cause her to be small, she may have more bad luck. Of all tapeworms, *Diphyllobothrium latum* sometimes selectively removes enough vitamin B_{12} to cause pernicious anemia.

THE PERFECT PARASITE

The ancestral tapeworm oozed from primeval seas along with some primitive chordate that had accidentally swallowed it. The animal intestine was its biological niche. To penetrate this paradise (a dark, warm refuge free of enemies, bathed in digested food), it overcame the problems associated with penetrating the highly acidic gastric juice in the stomach to lodge in the small intestine. There, floating in nutrients and protein-digesting enzymes, it found paradise. But there's a catch. Being made of proteins like all animals, how could it survive in a bath of protein-digesting enzymes? Over vast millennia, the tapeworm evolved an "antienzyme." When alive, it secretes this substance and remains intact. When it dies, it is digested along with dinner and soon disappears, adding its proteins to those of the meal.

But another threat arises. To win the game of chance that represents survival, it must produce uncountable eggs and broadcast them into the environment. It does this by living for a long time, sometimes for twenty years.

The tapeworm is a flat, featureless, macaroni-like organism without eyes, nose, heart, gut, brain, legs, etc. It is virtually a bag of sexual organs and eggs. Without a mouth, how does it eat? Without nose or eyes, how does it find food? Without arms or teeth, how does it capture food? One answer is that over time, it has lost those structures that are of negative survival value. What need for eyes in the darkness of the intestine? Eyes would only be vulnerable to infection and reduce the chance of survival. What need for a potentially troublesome gut when the worm is bathed in digested food? Those tapeworms that appeared without these organs had a better chance of survival and passing on their genes.

How, then, does a tapeworm eat? It has evolved a fuzzy surface covered with microvilli—microscopic absorptive projections that make the surface blotterlike. Energy-rich cells are interspersed among the epithelial cells comprising the walls of the microvilli. These pull molecules of the host's digested food across the surface of this fuzzy membrane (this also happens in the walls of the human gut).

An enigma arises: if the tapeworm eats an inordinate amount of the host's nutrients, it will harm the host, reducing its capacity to survive. So it absorbs little, removing few nutrients, its slow metabolism requiring little more energy than that needed to produce thousands of eggs. (Selective removal of vitamin B_{12} may be a rare "evolutionary mistake," like any harmful inherited condition in humans.)

Rarely does pain make a healthy host aware of the presence of even a huge tapeworm. Rarely does a tapeworm in the gut cause excessive hunger. Even more rarely does a tapeworm extract enough food to cause the host to lose weight. So little harm is caused by a relative of the gefilte fish tapeworm, the monstrous hundred-foot-long *Polygonophorus*, that a sperm whale can live to be one hundred years old or older with the tapeworm inside its gut for much of its life.

There was a monumental clash between the two titans of twentieth-century biology, Horace Stunkard, supreme parasitologist of his day, and the cigar-smoking saint of taxonomists, Libby Hyman.

The controversy: Is the tapeworm—literally a sack of eggs, virtually bereft of all but reproductive organs—primitive or advanced? Is the tapeworm, the featureless blob we see today, frozen in its primeval state? Or, despite its disreputable appearance, has it evolved

PLATE 10

A. GEFILTE FISH, BROAD TAPEWORM, *Diphyllobothrium latum.* A veritable monster; to 40 feet; averages 30 feet; longest human tapeworm. Sheds a million eggs a day. Scolex has two thin-walled elongate suckers; up to three thousand proglottids graduated as to maturity and size. Those nearest scolex are *immature;* inactive reproductive organs develop until they become *mature,* with developed sex organs. Branched dots are testes; curved tube, uterus with birth pore at top; dark "wings," ovary; no gut. They develop until they become *gravid—* filled with eggs; in some cases one hundred gravid proglottids release eggs daily through birth pore into gut, unlike all other human tapeworms that lack birth pore.

B. EGGS REACH WATER. "Door" (operculum) at top opens, releasing larva that becomes

C. CORACIDIUM, microscopic; round; swims with cilia. Six larval hooks not functional.

D. COPEPOD EATS CORACIDIUM that develops in body cavity into encysted procercoid. Minnow eats copepod, which in turn is eaten by

E. YELLOW PIKE, *Stizostedion vitreum,* second intermediate host. In fish, procercoid matures, becomes plerocercoid (sparganum) that migrates to muscles of fish until eaten by elderly lady.

F. PLEROCERCOID. To more than 3 inches; white, flat, featureless; resembles macaroni; tiny scolex embedded in anterior. When raw fish is tasted, the woman (accidental definitive host) is infected. Usual definitive host is bear.

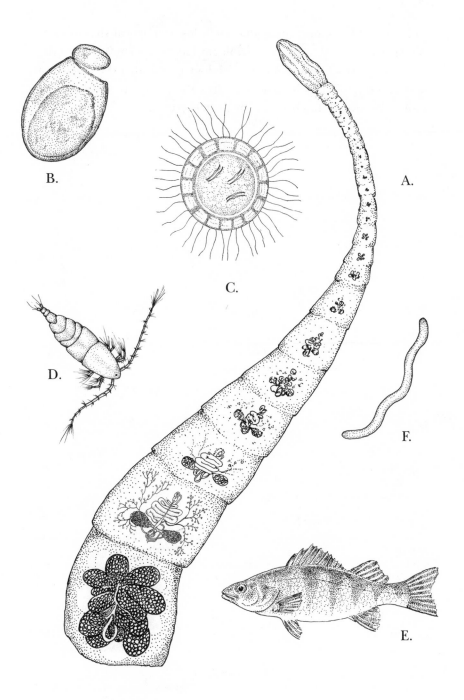

A.

B.

C.

D.

E.

F.

to be a highly sophisticated superblob, losing physical structures as time passed and adding new physiological features? The advent of the electron microscope and microphysiology reveals that the tapeworm is extraordinarily highly adapted to its habitat on a microscopic and molecular level. It has become the perfect parasite.

11

MOTHER ALWAYS WANTED ME TO BE A *REAL* DOCTOR

THE SECRETARY HANDED ME THE PHONE, a look of consternation on her face. Evidently this was some sort of emergency. "Hello? Is this Doctor Kaplan?"—a woman's voice bordering on hysteria. "Yes?" "Please come over, my husband . . . " she hesitated. "He is sitting on the toilet and afraid to move." "Why?" I asked. "He says that there is a white string hanging from his anus," she shrieked. "Please come over immediately!"

"I am not that kind of doctor, madam. You need a physician." Feeling inadequate, I tried to sound as compassionate as I could, and said softly, in my most professional voice, "But I can assure you that there is nothing to worry about." I had heard this said countless times on television. "It is nothing but a tapeworm, and they are harmless." "Harmless!" she bellowed, "how can a tapeworm be harmless?" I felt I owed her an explanation, but it was difficult for me to begin to explain the intricacies of host-parasite physiology to a hysterical person. I tried my best but had the feeling that I had failed. I gave her the telephone number of my physician and told her to call him immediately. I was sure he would send the patient to the emergency room, and that there would be a crisis when the husband had to get off the toilet.

THE PUZZLE OF THE CLOGGED GUT

The rice rat, *Oryzomys palustris*, seemed ordinary. Fresh from the Sherman trap, it had died overnight and was still warm. Its brown fur was matted with the urine and feces that accompany death. The post mortem was routine. I checked the lungs, abdominal cavity, and stomach, then the intestines. Wow! They were literally distended with dozens of tapeworms. "How come this ordinary-looking rat is not emaciated?" I asked myself. The small intestine was so clogged with tapeworms that it appeared nothing could squeeze through.

Evidently, the rules of parasite survival were broken by this rat tapeworm, *Hymenolepis diminuta.* It is illogical in an evolutionary and ecological sense for there to be a mass of parasites clogging the gut. Survival of the host may be impaired, threatening the parasite's survival. Why would a tapeworm, one of the most highly evolved parasites, endanger itself in this manner?

But how could it do otherwise, given the abundance of beetles—most of whom contain larval tapeworms. Eat a beetle, get a tapeworm . . . eat another beetle, get another tapeworm.

A rat running across the living room floor creates consternation. It is as if the members of the family standing on chairs know why they are so afraid. But they don't. They don't realize that the rat tapeworm can easily enter the human body and thrive in the human intestine. The flour beetle, *Tribolium confusum*, lives in the pantry, going through its life cycle in the flour container. Occasionally a beetle eats rat feces. A rat eats the beetle. Rat feces can contain hundreds of eggs. In each egg is a six-hooked tapeworm larva. When eaten by the beetle, the larva escapes its eggshell, burrows through the gut wall, and enters the beetle's body cavity, turning into a tiny-tailed gob of tissue with a headlike scolex buried inside. The clever tapeworm prevents female beetles from maturing sexually. With no energy used up in sexual development, more energy is available to sustain the parasite. When the beetle is eaten by a rat—or child—the larva burrows into the intestinal wall. After five or six days a fully formed immature tapeworm reenters the intestine and attaches itself to the inner wall, soon maturing into a reproducing adult twenty inches long. *Hymenolepis* is literally harmless to a healthy host. In fact it has been suggested to drop it from the ranks of parasites and call it a commensal.[14]

It is well that Mom becomes upset, for food can become contaminated by rat feces and directly infect the family without a beetle intermediate host. Rates of infection by the closely related *H. nana* are 1 percent in the southern United States, 9 percent in Argentina, and an astounding 97 percent in Moscow. Evidently Moscow has a rodent problem. *H. nana* is more common in humans because the worm can complete its whole life cycle in the human without needing an intermediate host. A study in Connecticut of rats being sold in pet shops revealed that one-third were infected with *H. nana*. Scary pets.

The puzzle of the rat's clogged gut stayed with me. To solve the problem required a large sample and many technicians to do the analysis. The solution? Use the fifteen students in my parasitology class as technicians. I waited until the students acquired skills with the microscope adequate enough to identify eggs in feces and do blood counts. Then I sent away for infected *Tribolium* beetles. Yes, there was a place that sold infected beetles.

Each student dissected a beetle, removed the larvae, and placed them in saline solution. Gloves on and teeth gritting with determination, pairs of students sucked up individual larvae into a pipette. One held the rat's mouth open, and, hands shaking with fear, the partner squirted a larva into its open mouth, bypassing the huge, scary incisors.

One of the experimental rats received one larva, another three, and so on until several rats received twelve or more larvae. Then we waited a month and checked each rat's feces. All contained eggs. It was time to sacrifice the rats and count the adult tapeworms in their guts. Results:

Rat fed one cysticercoid (larva): one adult tapeworm in gut
Rat fed three: three adults
Rat fed six: four adults
Rat fed nine: five adults
Rat fed twelve: five adults

PLATE 11

A. EGG OF RAT TAPEWORM, *Hymenolepis diminuta*. Contains six-hooked hexacanth embryo (onchosphere). Larval hooks disappear before maturity.

B. CONFUSED FLOUR BEETLE, *Tribolium confusum*. Tiny, 1/10–1/8 inch, flat, shiny, reddish brown. Can go through whole life cycle in flour container. Rat feces have odor that attracts beetles; eggs hatch in beetle gut and become cysticercoid larvae in beetle's body.

C. CYSTICERCOID. Tailed bladder with scolex inside. Larval hooks in tail soon disappear.

D. SQUIRTING CYSTICERCOIDS INTO RAT'S MOUTH. Infected beetle dissected under microscope, freeing cysticercoids. Mouth held open by one student; the other squirts. Rat unhappy, squirting difficult.

E. GRAVID PROGLOTTID of typical terrestrial tapeworm—pork tapeworm, *Taenia solium*. Branched uterus filled with thousands of eggs. No birth pore; proglottids must pass out in feces to be eaten by intermediate host, pig. Pork tapeworm atypical, cycle does not need intermediate host; humans can be infected by poorly cooked pork.

F. SCOLEX OF *Taenia solium*. Similar to rat tapeworm; single row of hooks and four suckers.

G. PROGLOTTIDS OF DOG TAPEWORM *Dipyllidium caninum*. To 1/2 inch long; ends tapered; usually attached in strings of four or five; resembles bracelet.

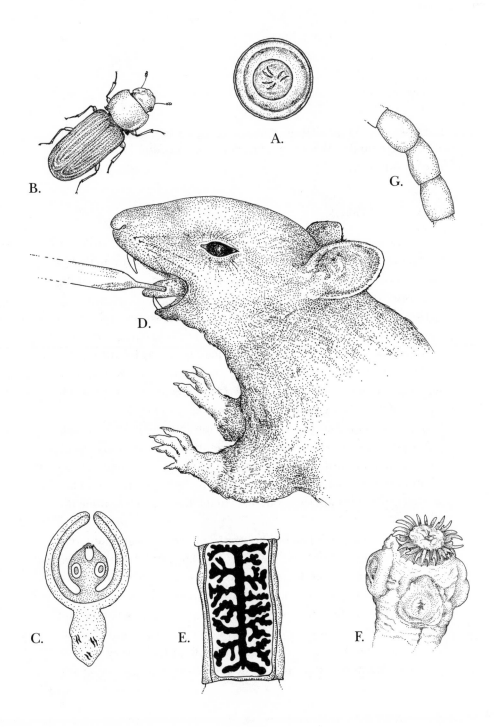

A.

B.

G.

D.

C.

E.

F.

Rat fed fifteen: seven adults
Rat fed eighteen: six adults

The tapeworms were regulating their population.

Over the years I repeated the experiment many times, and the results were always similar. What caused the massive infection I found in the rice rat still puzzles me.

DRAMATIC DOGS

The dog tapeworm, *Dipylidium caninum*, is pretty. The proglottids are tapered at each end into a lovely chain that I have seen imitated in jewelry. I sense that the jeweler and his patron are not aware that the "exclusive" design also exists in the bowel of a dog.

Dog tapeworms do not release eggs into the host's feces, and there is no egg-filled uterus. Instead, the eggs aggregate into clumps of ten or fifteen inside each mature proglottid. There is no way for the clumps to escape! Instead, when the eggs are ripe, *the proglottids crawl out of the dog's anus!* Sometimes they appear as writhing chains of four or five. Or worse, individual proglottids drop out and rapidly crawl away.

In this day of urban sanitation, the person with the inverted plastic bag or pooper scooper awaiting the excretory function of the beloved dog is sometimes in for a shock. Fido has the string! Dangling from its anus are egg-filled sacks that will fall to the ground and eventually rot. Not so the resistant eggs. But how do the tapeworms continue the cycle?

Enter the flea. Each flea on the dog lays dozens of eggs every day. These fall to the ground and hatch into grubby little larvae. The larvae eat any rotting stuff, including the feces—and tapeworm eggs—and pupate in a tiny cocoon, all the while harboring newly hatched larval tapeworms. One final molt and the flea becomes an adult, still retaining the larval tapeworms. It jumps on a dog (or another vertebrate) and bites. Like a vampire, it needs blood for its very survival. The flea sinks its mouthparts into the dog's skin. The dog feels pain.

It nips at the wound, eats the flea containing the larval tapeworm, and swallows it.

The life cycle is complete.

There was a product sold at the World's Fair in 1938 that was billed as a reducing pill. A scandal erupted when it was revealed that the active ingredient was a mass of scolexes of the human beef tapeworm, *Taenia saginata.*

The tapeworm pill didn't work when taken by healthy, robust, overweight people.* After all, how much food can a wafer-thin, ten-foot tapeworm eat? As we have shown, tapeworms regulate their populations and otherwise minimally intrude on the host's metabolism. A happy host is a long-lived host!

To see how far tapeworms will go to keep their hosts happy, consider a relative of the gefilte fish tapeworm, *Spirometra mansonoides.* The intermediate macaroni-like stage, when experimentally injected into mice, causes them to gain weight and become obese. No one knows why for sure. One informed guess is that the larva interferes with the metabolism of the mouse and damages the pituitary gland. It is possible that the tapeworm takes over the secretion of pituitary growth hormone and causes excessive tissue proliferation.

Tapeworms love a fat, happy host.

---

*The beef tapeworm, *T. saginata*, does cause symptoms in unhealthy people: abdominal pain, weakness, excessive eating, and loss of weight—but almost invariably in unhealthy people.

12

MISSUS MURPHY'S BABY

MISSUS MURPHY LOOKED OUT OF THE WINDOW at the dun-colored vastness of the Australian outback and sighed. Holding her bloated belly, she wondered out loud, "When will this damned pregnancy end?" She was disgusted. It was hard to waddle outside and face hours of intense heat and flies in the sheep-shearing shed. This was the day after she was due and she was worried. "Maybe the doctor miscalculated?"

A week later, a concerned doctor at the clinic told her, "There is no sign of activity—none of the preliminary contractions that happen before birth. Come back in a few days and we will take another look." Missus Murphy was getting nervous. By her count, this was the middle of her ninth month. She had had no contractions and no pain, and, lulled by the absence of symptoms, she waited . . . and waited. Finally, very scared, her husband drove her the forty miles in the '33 Chevy truck back to the clinic. The doctor took one look at her distended belly and rushed her off to the hospital for an emergency caesarean.

Thank god she was anesthetized. The surgeon's blade sliced a horizontal incision through the taut muscles at the peak of the moving mound that was her abdomen. The incision had hardly been made through the rock-hard, distended abdominal muscles when the birth began to take on a life of its own. To the doctor's horror, the slit became distended. A bloody reddish sphere was emerging from the confining abdominal organs as if it were being pushed from within. He realized that its huge intruding presence had created such pressure that it was freeing itself. He had never seen a newborn like this. "My god, she has just given birth to a bouncing baby ball." He felt bad for the nervous levity of his thought. "What is this thing? Maybe she has delivered the whole womb and the baby is inside?"

He raised his scalpel to free the baby and made the cut. Out burst a watery, grainy fluid that sprayed all over, covering his shirt and splattering his face. It tasted salty as it reached his mouth. The floor was covered with the fluid as it dripped from the now deflated fleshy sphere. It became silent in the room. No one knew what to do. The disciplined silence gave way to horrified gasps. The surgeon regained his equanimity. "Someone take this thing away and put it in alcohol and I will close up."

Suddenly the anesthetized patient began to gasp. Her heart rate began to rapidly increase; her throat was closing. She was going into anaphylactic shock. Her immune system was sensitized to the unknown fluid. She was dying. "Code Blue!" the surgeon shouted as he injected her with epinephrine.

A few days later, Missus Murphy was in tears. She could not believe that she had not been pregnant after all.

Later the mysterious sphere was measured. It was fifteen inches in diameter, contained about ten quarts of fluid, and weighed eighteen pounds. No one knew how to dispose of it. "It's the spawn of the devil, burn it." "Send it to the university, they'll find out what it is." Calm heads prevailed and a car was dispatched to the university medical school. It carried a garbage can containing the grisly specimen still soaked in alcohol.

The can made its way around the pathology department until someone scooped out some of the fluid from inside the deflated chamber and examined it under the microscope. Startled, he could not believe his eyes. Clearly visible in the bright circle of light was a scattering of tiny, whitish, beadlike balls. Barely visible inside each was a circle of hooks—the invaginated scolex of a tapeworm. The mystery was solved! Apparently the "baby" was a huge bag of tiny tapeworm tips. Dissection revealed that each of the many minute scolexes was armed with a vicious-looking circle of spines and four suckers. Missus

Murphy had waited too long. She had "given birth" to a hydatid cyst containing thousands of the smallest of all tapeworms, *Echinococcus granulosus*. The cyst grows so slowly that she must have carried the disgusting burden inside her for twenty years. Or it may have been older. She may have been infected when she was a child, in which case the worm may have been almost as old as she was.

Such a long "pregnancy" with such a horrible end.

EVOLUTION CONQUERS

Sometime in the long-distant past an aquatic animal slithered from the sea to invade a new, predator-free habitat—the land. It was safe! No enemies lurked behind a submerged rock, no huge aquatic monster would chase it. But an enemy lurked within. Hanging from its upper intestine was a white, seemingly segmented predatory beast, sucking nourishment from its intestinal juices—a tapeworm.

But in an evolutionary sense, the parasitic enemy was beaten! It was tied to an aquatic life cycle. It needed an aquatic intermediary, a copepod or minnow, to complete its life cycle. Without water there was no way to undergo larval development. This was an insurmountable dead end!

A BRIDGE OVER TROUBLING WATER

It had taken uncountable thousands of years for the tapeworm to overcome the defenses of its aquatic host. How could it make yet another transition to a host that had reached the protection of dry land? How could it find a terrestrial host when its larva needed to swim to reach its developmental haven? Without copepods there was no way to connect the larval swimmer to the first fishy intermediate host, a minnow. There would not be a large fish to eat the minnow. No definitive host to eat the large fish and complete the cycle.

But the tapeworm had an ally—time. Over more eons and innumerable evolutionary experiments, the frustrated tapeworm finally bridged the barrier. The aquatic tapeworm successfully entered the realm of terrestrial existence. Solution: the egg did not release a swimming larva, for there was no water.

The trick was to retain the larva in the egg. To break free of the egg was to die in the alien air. The infective larva had to wait, safe in its impervious capsule. It could not search for a host. It was trapped. Before the problem was solved, uncountable larvae died when their protective eggs dried in the desiccating air, buried in the host's dehydrated feces.

Over more time it came to pass that egg-laden feces were strewn about the landscape, advertising their presence by an almost-suffocating aroma. But the stench was like perfume to a class of animals that converged on the fecal mounds. A horde of insects, remotely related relatives of copepods (and as ubiquitous) devoured the feces and their parasitic contents. Over more time some of the eggs cracked open and freed the larvae in the insect gut. They burrowed into the insect's body and underwent a second larval stage. Then the insect was eaten by a carnivorous terrestrial mammal—not a fish. A mammalian host had eaten the insect. No water was needed. Thus was the terrestrial circle of survival completed.

THE TINIEST TAPEWORM

The villain of the piece, the mini-tapeworm *Echinococcus granulosus*, has only four parts: a headlike scolex, an immature proglottid, a mature proglottid, and an egg-bloated gravid proglottid. It is a quarter-inch long, and the comparison to the broad or gefilte fish tapeworm is like a human standing next to a skyscraper. Like most tapeworms, the adult attaches itself to the upper duodenal wall. Like most tapeworms, it is almost benign. Like most tapeworms, it is the pinnacle of the evolution of its kind. It is a perfect parasite. To be highly evolved is to be almost harmless—harmless enough so that the host survives for as long as possible. The parasite dictum: do not kill the golden egg.

The *E. granulosus* life cycle plays out like this: An infected dog defecates. The egg-laden feces contaminate a plant—usually grass, but sometimes a crop plant. Soon the feces rot and disappear. Not so their burden of eggs. The impervious eggs stick to the plant and are eaten by a herbivore, a sheep, or one of the vast population of rabbits that blight Australia—or a human who eats a contaminated plant. (A common source of infection occurs when a child is licked in the face by an infected dog that has "just groomed itself."[15])

The egg passes through the rabbit's (or human's) stomach protected by its shell. The shell cracks open in the intestine. The just-freed larva burrows through the intestinal wall, enters the bloodstream, and takes the rich red highway to anywhere in the body. It often gets off in the liver; less commonly in the lungs, spleen, brain, or bone marrow. Once at its final destination it encysts and reproduces asexually, budding off tiny tapeworm scolices in a bag—the hydatid cyst.

But the larva cannot become an adult in the herbivore (or human) intermediate host.

A dingo—the Australian wild dog—eats the rabbit. The larval tapeworm-laden cyst bursts open in the dingo's gut and thousands of tiny adults soon coat the inner intestines, causing little damage beyond sharing digested food with their canine host. The life cycle is complete.

But what if a human, like Missus Murphy, eats a plant anointed with eggs? She unfortunately becomes the herbivore intermediate host instead of the rabbit. The larva migrates through her gut wall and is transported by the blood, in this case, to the membranes of her abdominal cavity. Here is where the disaster occurs. Unlike any self-respecting larval tapeworm, the larva grows to form a sphere, a hydatid cyst. Its inner wall divides madly, producing vast numbers of tiny balls. Inside each ball are buried one or several scolices hanging by stalks into the fluid-filled chamber. The reproduction continues unfettered. Sometimes the scolices break free and form a layer at the bottom of the ball called hydatid sand. Gradually the sphere enlarges. It can become huge over a lifetime. Some hydatid cysts contain fifteen quarts of deadly fluid.

Humans are a dead end. The hydatid cyst goes on producing scolices and becomes gigantic—not the bite-size cyst in the rabbit eaten by the dingo.

The human host has three potential death-dealing disasters to deal with:

- The cyst gets so large it displaces vital organs, impairing their function.
- The larva, in its wandering throughout the body, leaves the bloodstream in the brain, lungs, or heart to disrupt their vital activities.
- The cyst breaks, releasing the fluid inside. A massive flood of this foreign substance becomes a huge challenge to the immune system resulting in anaphylactic shock—and death.

Dogs are the most serious threat to humans because they are our companions and they disseminate eggs. We insert ourselves into closely contained feeding relationships like sheep-dingo and warthog-lion and become accidentally infected.

By fondling infected dogs with eggs on their fur or eating egg-laden dog feces on plants, humans become unwitting substitutes for the herbivore in the cycle. Once trapped inside the false (human) intermediate host, the worms cannot reach maturity. They need the real carnivore. The larva is going nowhere.[16]

The disease is endemic and serious in sheep-raising areas of Australia and New Zealand. In Lebanon leather tanners have an unusually high incidence of infection. They seem to use dog feces in the tanning fluid. Imagine a nervous tanner biting his fingernails. Some tribes in Kenya become massively infected, betrayed by their penchant for stuffed dog intestines briefly scorched over a fire. Next time you eat haggis, stuffed derma, or even old-fashioned sausages, think of them.

Modern medicine has progressed a long way since this episode in rural Australia in the 1930s. Nowadays, the hydatid cyst is located during routine X-ray examinations or sonograms. Sometimes, when it is embedded in an inoperable area, it is first reduced by carefully inserting a long-needle syringe into the cyst. The hydatid fluid is carefully sucked out. If the surgeon makes a mistake and the fluid leaks internally, the patient can die of anaphylactic shock. Next, formalin (itself a carcinogen) is injected into the cyst, killing its contents and immobilizing the massively dividing inner layer of the cyst. The dead cyst deflates and is left in position or excised in a delicate operation.

One such organ needing a delicate operation is the eye. Thanks to modern surgery, the cure rate in infected eyes is high.

EASY SEX

Sexual behavior is stimulated by dark, warm places. (Ask me—I was the exhausted chaperone of a hundred coed college classes in the Caribbean.) Dark and warm, as in your gut. Add to that picture a hundred desperate "penises" seeking a hundred vaginas—*all in the same individual.* "Desperate" is the word, for if this well-endowed animal does not reproduce, its genes will not pass on and, without a mechanism for sexual survival, it will become extinct.

"Penis" is *not* the word, for that sophisticated organ has yet to appear. Stuck in an evolutionary sexual backwater, the needy tapeworm evolved an early, primitive structure to transport the sperms—the cirrus. Not as complex as a penis, the cirrus turns inside out (everts) like a children's party favor to provide the tube through which the sperm-laden ejaculate travels.

One would think insemination would be easy because each proglottid, in a body as long as a hundred proglottids, contains both male and female reproductive organs. Testes are sprinkled throughout the proglottid. The vagina leads to a chamber near the ovary where the sperms are stored. Each egg is fertilized by these stored sperms as it enters the long, tortuous uterus. This goes on day after day so that the uterus becomes engorged with thousands of eggs. So swollen with eggs is the uterus that it obscures the other organs, pushing them out of the way on its convoluted pathway to nowhere. No way out. There is no birth pore in a terrestrial tapeworm.

In the aquatic species the female disgorges hundreds of eggs every day—hundreds of thousands over a lifetime as long as fifteen years. All that is needed is one egg to successfully initiate the reproductive rite that continues the species through time.

But all tapeworms must rely on intermediate hosts to transfer their incipient life through a series of maturational stages, eventually to end up in the final (definitive) host. Infant becomes youth and youth becomes adult in a series of steps, each requiring a particular host for further development. To reach the definitive host, the mature worm must spill out thousands of eggs, in the evolutionary hope that they will be picked up by the proper intermediate host—an insect or small mammal (terrestrial tapeworm) or copepod or small fish (aquatic tapeworm).

The eggs must be acquired, not only by an insect or copepod, but by a particular species in most cases. This species specificity signifies that the tapeworm cycle has evolved over many thousands of years to a point of perfection, assuring that its genome gets passed on. This cycle is exquisitely adapted. Not all intermediate hosts will do because it takes complex genetic manipulation to direct the larva to burrow through the gut into the body cavity, where it can develop further in the evolutionary hope that this first intermediate host will be gobbled up by the second intermediate host—a minnow of a certain species in the case of water-based cycles, and a rodent in the case of terrestrial cycles.

Sex appears easy in the dark confines of the intestine. No sexual partner is required. The proglottid above or below will do. The cirrus everts; it enters the upper or lower neighbor's vagina. Sperms are released. Eggs are fertilized and pour into the host's gut, to reach the outside in the feces. Simple, eh? But there is a huge problem in this description.

The male part of the proglottid is part of the *same* tapeworm as the female part. Its sperms have the same genome as its eggs. No mixing of two unrelated partners' genes. No provision for the variation needed to make it possible to cope with changes in the environment (even the gut changes over thousands of years). Sexual reproduction and its genetic admixture is an evolutionary must. Hence the simplicity of the image presented above cannot occur. If gametes of the same animal unite, resulting eggs will be sterile, or if offspring develop, they will be sterile. Now the problem of mating becomes magnified a thousandfold. To provide the necessary variability, the tapeworm must line up alongside another of its species, cirrus to vagina, in the dark recesses of the intestinal lumen. How do they do this? No one knows. Just another of the mysteries of sex.

PLATE 12

A. This drawing by Gustave Klimt (1862–1918), *Pregnant Woman with Man*, clearly projects the pathos of Missus Murphy's dilemma. This woman is embracing her child with love even before it is born. Her head is bowed toward the new life inside her as if she wanted to encircle it; her whole concentration is directed toward it. Missus Murphy, tragically, went through what she thought was the whole gestation period anticipating her new child. Sadly, that was not to be.

At her feet are symbolic plants contaminated by eggs of *Echinococcus granulosus* shed by a dog or other carnivore, the natural definitive host of the tapeworm. Missus Murphy ate lettuce or another plant contaminated with dog feces.

B. HYDATID WORM, *Echinococcus granulosus*. Tiniest tapeworm, to ¼ inch; millions coat bowel wall. Four parts: scolex, immature, mature, gravid proglottids; gravid contains thousands of eggs that are shed in carnivore's feces.

C. *Echinococcus* EGG contains onchosphere larva with six larval hooks. When eaten by herbivore, onchosphere migrates to any organ, becoming hydatid cyst that enlarges over years. When Missus Murphy ate contaminated plant, she acted as herbivore.

D. INSIDE MISSUS MURPHY'S BELLY. Hydatid cyst filled with fluid; medium circle is daughter cyst; small circles are brood capsules. Protosolices (larvae) bud from inside of wall of each stage inside cyst; dots are free protoscolices, which become hydatid sand. When carnivore eats hydatid cyst, it gets a mouthful of thousands of protoscolices.

E. ENLARGED BROOD CAPSULE contains many protoscolices.

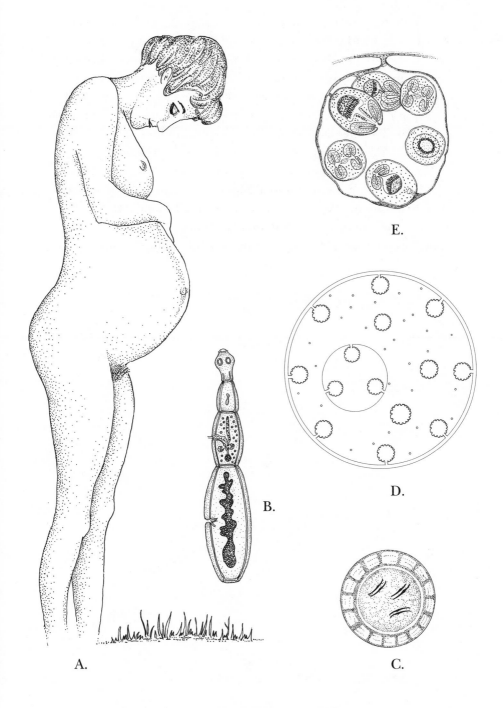

A.

B.

C.

D.

E.

But there must be some assurance that once found, the mates remain attached in connubial bliss so that the sperms can be transported. Not to fear, in some species once the cirrus turns inside out there is revealed a crown of spines (ouch) assuring permanent copulation. There is even a related fish fluke (monogenetic trematode), *Diplozoon paradoxum*, that remains in copula for so long that the two lovers fuse together like the mythical Hermaphroditus, so in love, so conjoined with his paramour, as to remain forever entwined.

Visualize the problem of aligning two adjacent worms so that a hundred cirri are inserted into a hundred vaginas. In some species that do not have a vagina, the problem is solved simply. "He" inserts his cirrus anywhere, plunging it into "her" body like a syringe. This is called hypodermic insemination. Saves a lot of time and effort.

13

THE DAY I FLUNKED
THE MACHO TEST

DEAD SOLDIERS (empty beer bottles) marched in serried ranks across the table to become by the end of the evening a veritable army. Beer-inspired lubricity lent animation to the intense conversation as young postdocs and younger graduate students tried their latest inspirations out on the director of the lab. Brilliant ideas were bandied about. New theories and clever hypotheses crashed against the wall of reason, the older and wiser chief scientist.

The throbbing drumbeat of the blues exacerbated my headache. I closed my eyes against the flickering neon lights. The cheap booze and beer-bottle smell made me nauseous. I longed to go home to my kids and wife. The ritual had become stifling. This was to be the cul-mination of my education—postdoctoral studies, the time when an aspiring scientist consolidates his research skills and prepares himself for a lifetime of scientific investigation. Here I was, sitting in a blowsy bar at 2 a.m., listening to the youthful inspirations of my colleagues. It was mandatory that we participate in this beer-inspired philosophical colloquy. How else would we hear what was going on in other labs?

Newly arrived, I was full of idealism. Not for long. My duties for the first several months included driving a careening station wagon down a dirt road in the middle of a swamp at night. My adviser and several graduate students were perched on the tailgate of the wagon, flashing a spotlight into the black waters edging the raised road. When an ominous pair of green-glowing dots reflected the spotlights—bang! bang! A fusillade of shots rang out. The glowing dots were the eyes of some unfortunate swamp animal. When the dots were extinguished, the mighty hunters leapt off the tailgate and plunged headlong into the midnight-black swamp. With a whoop and a holler they charged, oblivious of alligator and water moccasin, through the muddy water to bring back what they had bagged. This is called "collecting specimens." My macho mentor collected animal hosts in a manner that required maximum bravado.

Whatever they had killed was handed to me, the newcomer, for a midnight postmortem. After all, the animal had to be examined before any parasites in its gut became distorted by death, and before its ectoparasites (lice and ticks) felt the body temperature decline and deserted the sinking ship. Words rhyming with "sinking ship" came to mind as I faced the prospect of dissecting the bloated carcass of a muskrat, back at the lab.

Tapeworms and flukes (parasitic flatworms) extracted from the stinking gut were preserved, stained, and mounted on glass slides for further examination. Only then could I go home. This was my participation in a survey of the parasites of mammals of the United States.

This story does not end in glory—at least not for me. I had my chance at demonstrating my masculinity and failed.

PARTY TIME

There is a parasite, the oriental lung fluke, *Paragonimus westermanni*, that causes untold human suffering in Japan, China, and elsewhere in Asia. Its life cycle demonstrates the entwined relationship between the habits of parasite and host.

The story begins with your average Asian farming family. It is morning. Mom wakes up and yawns. She commences her ablutions. From under her bed she removes a lovely blue and white porcelain bowl. She squats and defecates into it. Sequentially, the members of the family anoint the thunder mug. (Use your imagination.) After breakfast, someone throws the contents of this glorified bed pan out the front door. They land in the rice paddy to become the only fertilizer poor farmers can afford.

The richly organic bottom of the rice paddy is ideal habitat for snails and crabs that voraciously eat the rotting rice stalks and dead animal matter. Hidden in human feces and spit are the myriad eggs of the lung fluke. Soon after contact with the water, they hatch, releasing ciliated larvae (miracidia) resembling microscopic hairy footballs. These spiral through the water until they accidentally come into contact with a snail.* They penetrate the skin of the snail, in the process being reduced to a ball of undifferentiated stem cells. These reproduce wildly in the liver of the snail, producing daughter cells by a process called polyembryony. The cells stick together, giving rise to many sausage-like sporocysts, each of which contains as many as twelve daughter sausages. These daughters have a mouth and can feed on the snail's liver tissue. They are called rediae. Two generations of rediae, mother and daughter, are produced. Inside the second (daughter sausage) generation of these translucent tubes, one can see movement. Wiggling tails and convulsive contractions announce maturation to the active, infective phase of the life cycle. Finally each daughter redia gives birth to dozens of motile cercariae that look like rubbery ovals with tiny tails. Miraculously, the few initial primordial cells injected by the miracidia into the snail have become hundreds, even thousands, of multicelled cercariae. They burrow through the body of the snail and crawl away. Cercariae are shed daily, like dandruff from an unwashed head.

Moving rapidly and randomly, these tiny-tailed crawlers accidentally find a crab feasting on rotting rice roots and human feces.† The crawling cercariae penetrate the crab's body and encyst around the heart. They go through further development, becoming metacercariae nestled into their protective chambers.

Then comes the party. A festive occasion in the Japanese rural countryside calls for quartered freshwater crab. The custom is to

---

*There is some evidence that in a few instances, miracidia are attracted to snails.

†In most species of flukes (trematodes), the cercaria uses a long tail to swim with. It looks like a huge sperm. *Paragonimus* cercariae, with tiny tails, are forced to crawl.

warm the crab pieces in hot sake and gulp them down. The heat of the wine does not kill the encysted worms; it serves only to stimulate them, and they writhe vigorously in the confines of their cyst walls, trying to break out. Eventually they escape and end up in the lungs of the human host, where they mature and lay their eggs.

THE MACHO TEST

The unsolved problem in the study of the oriental lung fluke is: how does the worm get from the farmer's mouth into his lungs?

Usually encysted larval parasitic flukes (metacercariae) escape from their imprisoning cysts in the small intestines and burrow through the wall, entering the blood. In many cases, larval flukes travel from the intestines through the liver, drift to the heart, mature in the lungs, then go back to the heart, the aorta, and back home to the intestines.

But my adviser thought a different, unusual pattern was employed by the *Paragonimus* larvae. A short cut. He believed they would move directly from the esophagus to the lung through the chest cavity.

He devised an experiment to test his hypothesis.

Live crayfish were dissected. In some, the microscope revealed telltale bubblelike cysts around the heart. These are characteristic of the American lung fluke, *Paragonimus kellicoti,* a parasite of weasels, otters, and cats—and the occasional human muskrat trapper who is fond of raw crayfish.

Chunks of thorax containing the cysts were placed on a stainless-steel tray. Its silvery reflections pierced the gloom of the basement animal room where the initial stage of the investigation was to take place. My "boss" opened a cage. There, pressed against the back wall, was the largest, meanest tomcat I had ever seen. One ear had been ripped off in a fight; scars crisscrossed its face. It was spitting and snarling. The boss (my mentor) said, "Open its mouth and I will shove this chunk of crayfish down its gullet." Obediently, my hand went forward and then—it stopped. It was as if it had a life of its own. I couldn't get it to reach in any further toward the furious cat. My boss saw that look in my eye and with disgust said, "Take these forceps and stuff the chunk into the cat's throat while *I* keep its mouth open."

I had flunked the macho test. My wayward hand would not do its duty.

I found out later why the cat was so furious. Just before this episode it had been injected with a harmless blue dye. It was literally transformed into a blueblood. Its intensely blue-stained blood made the whites of its eyes and tip of its nose bluish.

An hour after the feeding, the cat was sacrificed. We examined the esophagus. There, visible to the naked eye, were tiny blue dots, microscopic holes made visible by the dye-tinged blood that was escaping from wounds made by the parasites as they burrowed through the pinkish esophageal walls. The blue color indicated that the worms had excysted in the esophagus almost immediately upon being swallowed and had penetrated into the thoracic cavity, going directly to the lungs. *My adviser's hypothesis was valid!* He had discovered the unique method by which the parasite was able to enter the lungs.

We examined the lungs of another cat that had been infected a month before. Distinct brown rings stained the coral-pink lungs. We dissected a brownish area. It contained two fat, half-inch-long adult worms aligned front to back. Each brown stain was a halo of thousands of eggs produced by the encysted adults. We learned from this:

- Animals will go to any lengths to perform sexual reproduction. Even though digenetic trematodes (parasitic flukes) are hermaphrodites—even though they contain both sexes—they theoretically do not self-fertilize. A genetic imperative necessitates that they search the dark recesses of the lungs to find a mate, enabling them to cross-fertilize. Only then does the mixing of genes make possible the variability that is the stuff of evolution.
- The eggs must get to the outside to carry on the genetic blueprint that constitutes the species. How do they escape from the lungs? It has been shown that the eggs travel up to the throat by means of the very defense mechanism that is designed to remove contaminants like dust particles from the lungs. Microscopic hairs (cilia) beat upwards, carrying offending particles (and eggs) to the throat, where they are either swallowed or spit out. In this instance many of the eggs that reach the throat irritate the

PLATE 13

A. How would you like to place a crayfish abdomen with metacercariae into this cat's mouth?

B. EGG CONTAINING GERM CELLS. Note operculum (door) at top.

C. EGG HATCHES IN WATER. Releases ciliated miracidium larva (hairy football).

D. MIRACIDIUM penetrates first intermediate host, snail, entering its liver.

E. MIRACIDIUM CELLS BECOME SPOROCYSTS in snail liver. They have no mouth, gut; and give birth to

F. REDIA. Also in liver; has mouth, pharynx; able to feed on liver; contains many cercariae that leave via birth pore.

G. CERCARIA burrows out of snail; has unusually tiny tail and crawls on bottom. Has anterior stylet and penetrating gland cells that help cercaria enter joints of Chinese mitten crab.

H. CHINESE MITTEN CRAB, *Eriochier sinensis*. To 3 inches across, light brown, hairy claws with white tips. Second intermediate host. Encysted metacercaria in body becomes adult in human when poorly cooked crab is eaten.

I. ORIENTAL LUNG FLUKE *Paragonimus westermani*. To 1/2 inch long; thick, reddish brown. Y extending from anterior sucker is gut; dots on sides are yolk glands. Black ovary near ventral sucker; tube from ovary is uterus. Paired black testes.

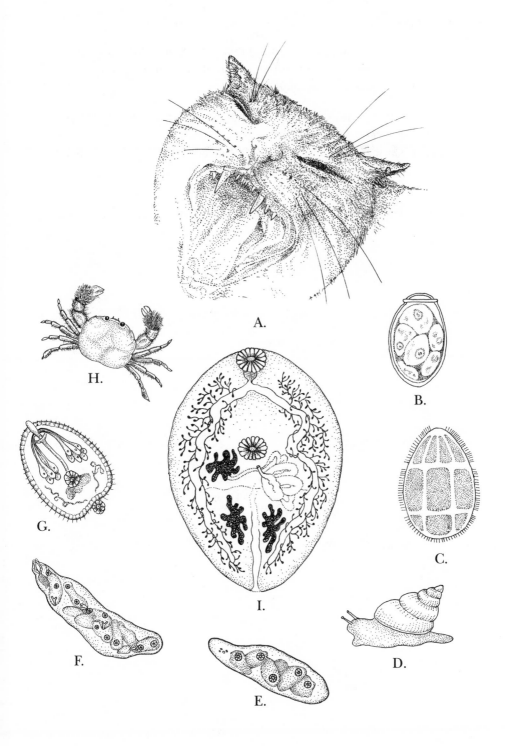

A.

B.

H.

G.

I.

C.

F.

E.

D.

farmer, and he spits or coughs them out. Some of the thousands of eggs are swallowed and are released in the usual manner with the feces—a two-pronged attack.

The egg-rich saliva and feces reach the muddy water. Another Japanese feast begins. Crab chunks are warmed in sake and the delicacy is served. Cups are raised. A toast. Good health to all. The life cycle of the oriental lung fluke is renewed.

14

THE BIBLICAL PLAGUES

THE LITTLE BOY AWAKENS in the stifling heat of the desert sun. He stretches and walks out of the shade of his straw-and-mud brick house into the glare of the treeless Sahara. Ambling sleepily, he wanders to the nearby irrigation canal and pees into it. The beginning of another mundane day. But this scenario of morning's nascence is different. There is a blood in his urine, making it appear orange. He watches as the colorful stream of urine merges with the slowly flowing current. This boy is completing an ancient biological ritual—the passing of a deadly parasite.

Afterwards, he and the other boys plunge into the cool water in an explosion of screams and laughter. Hidden in the rich soil on the bottom is a population of snails larger than all of Egypt's huge populace. Inside the liver of many of the snails, a parasite is multiplying. Every day each infected snail sheds a horde of malicious fork-tailed, spermlike swimmers. They cannot see, but they can sense shadows. The miniature monsters undulate through the water, tails flapping. One senses a shadow. It approaches, and with all its strength it hurls itself into the dark mass of an underwater rock—and dies. Of the millions that perish, one or two find a human. Using a skin-dissolving enzyme, the survivor burrows its way into the skin and finds the blood-rich dermis, entering a capillary. It is home!*

It is taken by a series of biochemical signals to the lungs, heart, aorta, and liver, maturing all the while during its journey. In the liver the adult worm seeks a mate. But the liver is not the end of its travels. Eventually it reaches its permanent haven. Following a genetically designated map, the mated pair arrives at the veins above the urinary bladder. There the couple copulates, producing, in a long lifetime, thousands of spiny eggs that burrow into the bladder. The victim spreads the disease whenever and wherever he releases his blood-tinged urine.

SNAILS AND ENTRAILS

The mud of the irrigation canals is riven with the slime tracks of snails whose abundance seems to rival the number of stars in the sky. Two or three out of every hundred contain a writhing mass of flaccid, featureless, saclike parasites (similar to the "sausages" of the previous chapter), mother sporocysts, their translucent walls revealing a generation of clones. When ripe, identical daughter sporocysts burst out of their mother sac. This new generation feeds by absorbing molecules of food through the skin. The snail's liver is a rich source of nourishment. The energy obtained from their parasitic banquet permits production of yet another generation. These are not the fecund sausages of the parent generations. These are the disseminators. Each has a forked tail. Each contains penetrating glands that can release a tissue-digesting enzyme similar to hyaluronidase (the enzyme that allows the sperm to penetrate the protein/mucus plug at the mouth of a woman's cervix). This third generation, the cercaria, forsakes the richly nurturing snail liver, burrows through the snail's body, and bursts out into the water. Forked tail undulating, it finds a human and penetrates the skin. The tail snaps off. Once inside, the cercaria, now

---

*Recent estimates suggest that the concentration of cercariae in irrigation canals in Egypt reaches 1.1 cercariae per 10 gallons of water, making it extremely likely that infection will result from swimming or wading.[17]

a mass of undifferentiated stem cells, drifts down the bloodstream to the human host's liver.

Blood-engorged channels flow sluggishly through the liver, making it an ideal haven for further development. Nourished by blood and liver tissue, the worm goes through adolescence. Finally it becomes a sexually mature adult.

Unlike all other flukes, this sophisticated super fluke has discovered an advanced form of sexual reproduction—two separate sexes. Instead of the usual cross-fertilization of two hermaphrodites, male and female have found their mates in the dark inner recesses of the liver. It is the male that is the seducer. He produces that universal elixir of sex, a pheromone. She follows the perfumed path to his voluminous embrace. But what is the nature of this embrace? Not with welcoming arms, for worms have no arms. Instead he has evolved a lengthwise fold, the gynecophoral canal, running almost the full length of his body. The seduction over, the female nestles into the canal and they leave for their honeymoon. But an obstacle arises that will test his mettle. The male has to carry his bride down the huge vein draining the urinary bladder and the rest of the posterior organs. But the blood is flowing upward, back toward the liver and heart. The male, carrying his bride, takes this in stride. He is so virile as to be able to swim against the flow. Instinctively, he enters one of the tiny veins draining the urinary bladder and takes up residence there. If he does not choose the correct tributary, the couple perishes. The preprogrammed path that carries them home is buried in his genes.

Once home, they set up housekeeping and mate—not once, but forever.*

BLOODY URINE

Unique to this species, *Schistosoma haematobium*, the adults live in the blood vessels above the urinary bladder. The hundreds of eggs released by the female each day have minute terminal spines. These spines were originally thought to help penetrate the tissues of the urinary bladder.

---

*Up to thirty years.

Now it is thought that, as in *Leishmania*, the invader tricks the host, using the body's immune system as its ally. The eggs cause an inflammation. White blood cells, responding to the irritants, carry the eggs through the bladder wall, riddling it with microscopic bleeding holes. Although the lesions are minute, their numbers cause considerable bleeding into the urine, often a pint a day. The resulting anemia is one of the symptoms of urinary schistosomiasis (bilharziasis).

A relatively uncommon symptom (common in related species) in untreated, severely infected children is a grotesque enlargement of the abdomen caused by the swollen liver and spleen (and a fluid symptomatic of pathology called ascites). The infected child becomes an adult, possibly with multiple schistosome infections. He is short because his growth was stunted by rigors of this disease. He has cirrhosis of the liver, scars caused by the passage of innumerable worms undergoing maturation there. And he has the characteristic lassitude attributed to the Egyptian peasant, due to anemia.

Schistosomiasis is the second most devastating parasitic disease afflicting humanity.[17] Its ubiquitous distribution spreads the ravages of the disease all over the world. There is the odd chance you can get schistosomiasis in Puerto Rico and other Caribbean islands by bathing in slowly flowing freshwater streams.*

Humans have been afflicted for thousands of years, as witnessed by the discovery of the characteristic cirrhosis of the liver found when Egyptian mummies were examined. Modern-day descendants of the mummy are becoming infected in increasing numbers due to the expansion of irrigation ditches coming from the huge lake resulting from the erection of the Aswan Dam to tame the Nile River. The plan was to make the desert bloom, but there were unforeseen disastrous ramifications. The explosion of irrigation ditches radiating from the

---

*At least 200 million cases of human schistosomiasis exist in the world. In China alone it has been estimated that at least 15 million cases exist. In lower Egypt an incidence of 60 percent has been estimated in most areas, with an incidence of 90 percent in some localities. In New York, Chicago, and Philadelphia the disease exists in "considerable" numbers. Fortunately the proper snail intermediate hosts cannot survive in temperate climes, and consequently the parasite's life cycle has not been established. But consider global warming.

lake extended the range of the parasite-carrying snail, increasing the infection rate a hundredfold. But the disaster had more profound unanticipated catastrophic effects on Egypt's long-suffering population.

MODERN-DAY PLAGUES

The building of the Aswan Dam produced devastation not seen in Egypt since the biblical plagues. It was as if God again visited his wrath upon Pharaoh. Instead of frogs, blood, vermin, and locusts, human intervention spawned these modern plagues:

Plague 1. The dam produced huge Lake Nasser. It regulates the level of the Nile and prevents its annual flood. The result is that the fertility of the soil around the Nile declined. Before the dam, one could fly over the Nile and see a broad swath of green decorating each bank. This voluptuous green band, brought to greater intensity by its stark contrast to the Sahara's sands, was the lifeblood of Egypt, a six-mile-wide Garden of Eden. It owed its existence to massive fertilization every year by annual floods that covered the fields with fertile river silt. But the dam tamed the river to stop the often-terrible inundation of the land. Absent the annual floods and their rich muck, the soil became less and less fertile. Corn, barley, and rice stalks became shorter and shorter. Their heads contained fewer seeds to grind into bread and nutritious porridge. The green band of life became pale. The answer to this modern plague? Buy fertilizer from the West, thus creating a negative trade balance with the United States and Europe. Those peasants who couldn't afford the subsidized price of the fertilizer perished.

Plague 2. May be the authentic, original plague that killed Egypt's first-born. It is commonly accepted that black (bubonic) plague originated in Egypt. Indeed, it is transferred by the oriental (Egyptian) rat flea, *Xenopsylla cheopis*, named after an early pharaoh, Cheops. In 1799 an astounding 92 percent of Napoleon's army died from black plague during his Egyptian campaign.

Modern Egypt is more crowded than ever, and rats flourish there. Black plague is a disease of rats carried by fleas to humans.

When the rat epidemic kills a huge proportion of rats, the fleas are forced to seek blood from other sources—incidentally humans.

Rat habitat is extended by the expansion of storehouses for produce from fields irrigated with Lake Nasser's waters. Every year there are a few cases of plague in Cairo.

The full fury of the black plague awaits.

Plague 3. Before the dam, the Nile's annual floods carried mineral-rich effluent into the Mediterranean Sea, fertilizing the water at the mouth of the river. This produced a massive bloom of microscopic plantlike cells, phytoplankton, forming a rich broth for herbivorous sardines that flourished until the annual flood stopped. No flood, no fertilizer for the phytoplankton. No phytoplankton, no sardines.

Before the dam, spring would bring huge schools of sardines. An army of fishermen harvested them, providing the impoverished Egyptian population with much-needed protein. When the floods stopped, the protein stopped and malnutrition prevailed.

The ancients saw that the annual floods brought life anew to the parched land. This concept of renewal was the foundation of ancient Egyptian religion. The pharaohs believed that their life would be renewed just as the river renewed life along its shores. They built elaborate tombs to wait until it was time to cross the river of death and back—the pyramids. The presence of these pyramids today provides a timeless bridge from biblical plagues to modern ones.

Plague 4. As the lake level rose, thousands of villagers were displaced, their homes inundated. Ancient temples, such as the one at Abu Simbel, were threatened. A mighty United Nations effort resulted in the removal of the temple and its reconstruction, stone by marked stone, on a cliff above the projected height of the lake. But the villagers were dispersed to the winds.

Plague 5. The modern greening of the desert brought death to the firstborn (and other) children through the expansion of the range of the schistosome-carrying snail into the tremendous web of irrigation canals. Before the completion of the dam, one in ten Egyptian children became infected with schistosomiasis, often fatally. A sevenfold increase in the infection rate was recorded four years after the completion of the dam in the surrounding population. In the lake above the dam, 76 percent of the fishermen became infected.

Plague 6. Like the sixth biblical plague, boils, there is a modern-day affliction, cancer. Daily penetration of the wall of the urinary bladder by hundreds of schistosome eggs brings more than anemia. Cellular and tissue changes in the walls, hyperplasia and hypertrophy (cellular proliferation and enlargement), and granulosis (pigmented particles in the afflicted cells) resemble, and can lead to, a variety of cancers.

There is strong evidence that bladder cancer is associated with schistosomiasis.

Plague 7. The Egyptians sought to destroy the Israelites in the desert again. In 1957 thousands of chariots (tanks) crossed the "Sea of Reeds" (Suez Canal) at the forefront of massed Egyptian armies. As before, the modern-day Israelites dramatically survived. Why? Some say it was because the fellaheen (peasant-soldiers) were so debilitated from schistosomiasis that they could not fight well.* This issue has been debated ever since the war.

But there is clear evidence that the Japanese schistosome, *S. japonicum*, prevented a war between mainland China and Taiwan in 1958–59. Alarmed by naked threats from Communist China, the United States sent destroyers to patrol the Straits of Taiwan, declaring America's intention to prevent Chinese armies from crossing. Despite the threat of American intervention, the Communist Chinese shelled the stepping-stone islands of Quemoy and Matsu in preparation for an impending invasion. A catastrophe of biblical proportions was brewing.

---

*Other explanations include the archaic leadership structure of the Egyptian Army; inability of poorly educated soldiers to use sophisticated modern weapons; and the superiority of equipment and tactics of the Israeli Army.

PLATE 14

A. Famous 24-inch-high statue in Brussels, *Manneken Pis*, by Jerome Duquesnoy (1619), symbolizes thousands of African children releasing blood-tinged urine every day. Urine contains eggs of *Schistosoma haematobium* that release miracidia to infect snails. Infected snails shed cercariae (D, E) that penetrate the skin of anyone in contact with contaminated water.

B. URINARY BLOOD FLUKE, *Schistosoma haematobium.* To 5/8 inch; male thick, has ventral gynecophoral canal in which slender female resides in permanent copula; female to 30 percent longer, cylindrical, posterior brown with eggs. Skin of male spiny.

C. SCHISTOSOME EGGS. *S. haematobium* elliptical; has terminal spine. *S. japonicum* spheroidal; has tiny lateral knob-like spine. *S. mansoni* elliptical, has long lateral spine; enlarged to show penetrating glands of miracidium inside that are used to burrow into snail. All egg sizes similar, all microscopic.

D. SNAIL SHEDDING CERCARIAE, *Biomphalaria* spp. Shiny cocoa brown, to 5/8 inch; intermediate host of *S. mansoni.* Can release a thousand cercariae a day. If ten thousand snails at 3 percent infection rate live near a village, three million cercariae greet swimmers, washerwomen, fishermen, every day (cercariae enlarged).

E. SCHISTOSOME CERCARIA, *Schistosoma haematobium.* Microscopic; swims with forked tail; note penetrating glands that secrete enzymes; can penetrate skin in 3 minutes; attracted by skin lipids, warmth. Must find host within hours or weakens and dies.

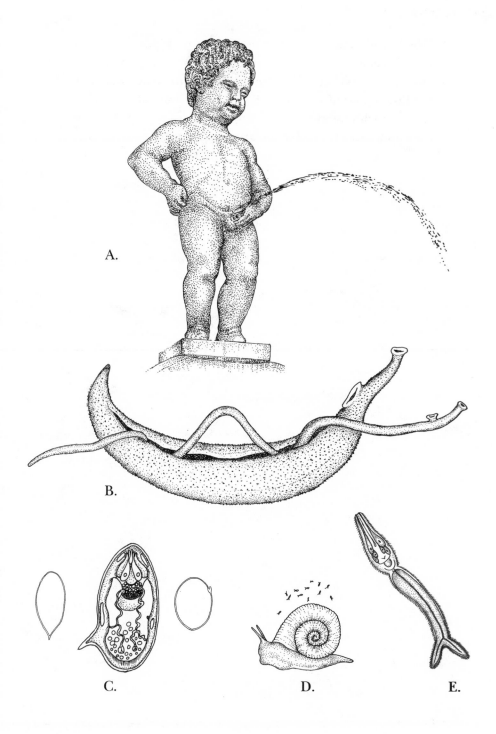

A.

B.

C.

D.

E.

The Chinese massed their armies. It took their primitive transportation system a year to accumulate the million soldiers necessary to ensure the success of the invasion. In the meantime the summer heat caused thousands of soldiers to bathe in the river. Many became infected with schistosomiasis.

The invasion was called off on account of sickness.

15

ALLEY CATS AND SEAGULLS

ALLEY CATS IN BROOKLYN were a persecuted lot. No wonder they howled, as if in pain, deep into the night. We street urchins hounded them unmercifully. My friend, Geno, and two buddies stealthily stalked their prey, a raunchy, scarred tomcat, with BB guns at the ready. This was not a lion-hunting safari, it was three teenagers creeping down an alley between two houses. But it was as dangerous as the African plains.

Geno and the rest of the safari penetrated deep into the alley, deep enough to be seen from the kitchen window. No cat. The intrepid hunters retreated to the street.

At this point, Mr. Umbriago, owner of the house, appeared at his front door and hurled himself down the stoop, roaring curses in Italian. As he leapt down the stairs, he drew his belt like a sword, producing, in my teenage mind, an image of a vengeful father in an Italian opera.

He lunged at Geno, grabbed him by the collar, and beat him mercilessly with his belt, uttering in Italian what I now believe to be oaths about sullied honor. It seems that Mr. Umbriago's plump, hairy, virgin daughter had reported a Peeping Tom peering at her through her bedroom window, and Mr. Umbriago had mistaken Geno for the perp.

I was clearly innocent, having been hanging out in the street while the invasion of privacy occurred. I waded into the imbroglio and pointed out to Mr. Umbriago that Geno was just trying to shoot a cat, not invade the privacy of his mustachioed daughter. By the time my protestations in unfamiliar English penetrated Mr. Umbriago's enraged psyche, Geno was a quivering hulk blubbering in the middle of the street. This trauma affected me profoundly. I never wanted to shoot an animal after that moment, which is a disadvantage for a parasite taxonomist.

HIRED GUNS

But hired guns were available for the price of a beer. Every year the culmination of the parasitology course was the necropsy of a seagull. These animals are remarkable for the variety of internal parasites inhabiting their guts and blood vessels. It is not their penchant for garbage that makes them vulnerable; the infections are long-term relationships between host and parasite evolved over millennia from the bird's not-so-particular eating habits.

Why is the corpse of a seagull so much more useful than any other corpse? Because the seagull is the most important model organism for studying schistosome parasites.

Schistosomes inflict their terrible toll on the lives of millions of humans by producing eggs that penetrate the walls of the urinary bladder or large intestine, causing severe immunological consequences, secondary infections, and infections of the liver, damaging this vital organ. Consequently, it is particularly important for future physicians and researchers to see them hidden in blood vessels just above the bladder or bowel.

Few American physicians are trained in parasitology, and fewer are able to diagnose schistosomiasis. When a tourist returns from Africa, South America, and many other parts of the tropical world* and exhibits anemia or a high eosinophil (a kind of white blood cell) count, the physician diagnoses the condition as an allergy and prescribes an antihistamine when an anthelmintic is necessary. The physician

---

*Two other species of schistosomes infect humans, *S. japonicum* and *S. mansoni*. Both are located above the bowel, and their eggs are found in feces rather than urine.

doesn't think to have the patient's feces examined for schistosome eggs. The debilitating condition may last for years.

No inaccurate diagnoses from *my* students! The seagulls have contributed to learning how to perform a necropsy on a "large" animal, and the students have seen schistosomes in situ. The awful smell of the opened abdominal cavity, the sight of the tangle of blood vessels in which the clotlike haven of the worm is hardly discernible, will stay with the students always.

All of this makes the sacrifice of a few gulls mandatory for a class of parasitology students.

KILLERS ON THE LOOSE

Early in the morning of the last lab of the year, I and my hired guns— my colleague and his son, duck hunters both—sally forth in a duck boat, the mandated variety of scientific collecting licenses sprouting from every pocket, to shoot a maximum of eight seagulls.

We approach a huge hill of garbage looming above the canal. Seepage from its mountainous flanks oozes into the water. This gigantic mound of garbage is paradise for gulls, and hordes arise from their disgusting meals at our approach. One makes the mistake of flying over the boat. Both mighty hunters follow it with the blued barrels of their 12-gauge shotguns and let loose a barrage at the high-flying bird. The ever-widening circle of small steel pellets hurtles toward the bird. It shudders when the pellets hit—and flies on. Seagulls are so tough that they often defy the powerful shotguns. Finally, one comes in low and across the beam of the boat. Boom! It drops to the water like a stone after the fusillade. I pick it up with a net. The guys get good at their job, and even high-flying birds come somersaulting down until we have our eight-bird limit.

Each pair of students is issued a gull. Cooperation is necessary else the lab would be inundated with the feathers that the students pull from the birds' breasts. (I assure them that, if all else fails, they can always get a job as a chicken-flicker).

Occasional feathers float through the air, dense with excited exclamations from the students. In the lab next door, students are silently hunched over their microscopes counting chromosomes. But here amid animated murmuring, a pair of students calls out to another to show them the filtered contents of the duodenum, a confused mass of yet-unidentified, feebly writhing parasites. Several species of small digenetic trematodes become identifiable in the dish under the microscope.

Visible to the naked eye are long, seemingly segmented white worms, each tipped with a tiny knob—tapeworms! This is the payoff. Long lectures abstractly cataloging exotic wormy organisms have come to life. Here, at the bottom of a dish, are huge, flat, fleshy tapeworms. At their anterior ends are the flabby-edged, long, oval suckers of aquatic tapeworms, infecting fish-eating animals including little old women.

A knot of students coalesces excitedly around a pair at the back of the lab. Their gull has been split stem to stern. The required midsagittal incision reveals the red-hued peritoneal cavity. They have followed the mighty inferior vena cava and are looking at its last skinny branch. A barely distinguishable bulge is the object that has drawn their attention. "That's just a blood clot," someone says. But the owners of the bird are obstinate. One holds the vessel with a pair of forceps while the other cuts out a segment containing the "clot." A dissecting needle is used for the final manipulation. The murmurs grow to a mini-roar. There, lying in the torn filmy layers of its protective blood vessel, is a clearly visible entwined schistosome couple, the huge male embracing its narrower mate in the protective folds of a groove running almost the length of his body. The anterior end of his mate protrudes to the length of her ventral sucker. She is a beauty!

The students run back to their birds, eager to find their own schistosomes. Several of the other gulls are infected, and excited exclamations attract yet other appreciative crowds.

These are bird schistosomes, specific to seagulls. They cannot survive in other kinds of birds or mammals.[18]

KING OF MUD FLATS

Change scene to mud flats covered with hordes of innocuous-looking mud snails, *Ilyanassa obsoleta*. The lowly mud snail, belying its mundane, grungy appearance, is extraordinary. Unlike any other snail, it is host to five, yes five, trematode larval stages that obtain nourishment from the richly energetic liver (hepatopancreas). In fact, when dissected under the microscope, the liver appears to be so filled with writhing, sausage-shaped early larval stages of the parasites that one cannot understand how the snails survive. But they do, even when 50 percent of the liver tissue is replaced by the "sausages."

One of the five species of digenetic trematodes found in the liver of the snail is the schistosome blood parasite, *Austrobilharzia variglandis*. Its life cycle is more direct than the usual two-intermediate-host process (see chapter 13). A sausage-shaped mother sporocyst produces a sausage-daughter sporocyst in the snail's liver. She gives birth to fork-tailed, eyed cercariae. A cercaria swims to the leg of a seagull and penetrates through the skin, leaving its tail behind. Thus one of the daunting elements of chance that jeopardize the completion of the ordinary two-intermediate-host trematode life cycle is eliminated— only one intermediate host, the snail, is necessary.

HERBIE AND THE ITCHY RASH

In the 1940s the world war hung over our lives. A dull roar overhead brought all eyes skyward, as echelons of warplanes flew in their in V-shaped goose-flocks toward Europe. My Brooklyn buddies and I were too young to fight. Much of our time was spent at Plum Beach. This wild expanse of sandy dunes and extensive intertidal zone had a dual use. By day it was heaven for nerdy little kids who wallowed in the mud of extensive flats exposed at low tide.

At night, it had a more lurid use. Due to Brooklyn's teeming population, there was scarcely room to escape from view. Privacy was sought after relentlessly. The parking lot at Plum Beach was "private," row upon row of cars notwithstanding.

A summer's sultry nights would stimulate the surging flow of testosterone in a teenager's blood and he would, with trembling earnestness, ask his sweetheart, "Do you want to go to the submarine races?" To the uninitiated this seems preposterous. Why would anyone go, in the dark of night, to watch submerged submarines race? But to the blushing young lady this signified something else. It was an invitation to make love in the front seat of a 1941 Plymouth or DeSoto sedan.

But we were too young for those shenanigans and confined our activities to daytime romping in the shallows. Myriad animals were to be found in the muddy sand and the tide pools. Hermit crabs would make their drunken way down a tiny rivulet. Sifting the sand would uncover worms galore. But by far the most common inhabitant was the mud snail. The flats from a distance appeared to be a dun-colored, undulating muddy desert, pockmarked with tide pools and curious-looking dots. On closer viewing, the dots became snails. Massive numbers of these animals surged across the flats, their siphons held aloft like an elephant sensing the air with his curved trunk. These inch-long snails were, in their intertidal world, like vast herds of elephants. They rushed to and fro, at a snail's pace, searching for carrion. When something died and began to decompose, the enticing odor of rotting flesh caused a slow-motion feeding frenzy. Mud snails slimed their way toward the stink. They converged on it, piling up on one another into a mound, jockeying for position, obscuring the corpse in a writhing mass of muddy shells and extended bodies.

The snails, like vultures, had a role in the cycle of life and death. They were nature's garbage men. Nothing rotted on that mud flat. The thousands of snails would see to that. They ate, among other things, the ubiquitous parasite egg-laden seagull feces.

HERBIE'S RASH

My friend Herbie's fascination with mud snails was unbounded. He would pile them up in huge squirming mounds, race them in tide pools, and otherwise find creative uses for these uncountable, innocuous-appearing animals. One day he came down with a fiercely itching rash. Blisters and lines crisscrossed his hands and arms. They itched so much he cried and rasped at his skin so fiercely that his mother took him to the doctor, who hadn't the slightest idea of what had caused the rash but treated it symptomatically. It soon went away. But it was apparent that the rash was associated with the mud flats and Herbie wasn't allowed to play with us at Plum Beach anymore.

Remarkably, deep in the heart of Brooklyn, Herbie had contracted the world's most terrible trematode disease, schistosomiasis. But this was a bird schistosome, specific to seagulls and a few other species of carnivorous shore birds.

Host specificity has two contradictory aspects:

- It provides a precise environment, in this case tributaries of the lower mesenteric veins of a seagull. This narrowly defined habitat frees the worm from competition with intestinal parasites and provides the richest of all parasite paradises, the oxygen- and protein-rich blood. On the surface this seems to provide perfect evolutionary logic: adapt ever more precisely to the host over time so that the worm is so well adjusted that it is almost part of the host's body and there is an ever-decreasing chance of being recognized by the host's defensive mechanisms.

- From an opposing view, precise adaptation narrows the potential for spreading the parasite's genes throughout a larger group of hosts. In an evolutionary sense, precise adaptation confines the worm to a specific habitat (the blood of the seagull) that might change over time, threatening the existence of that parasite species. Suppose seagulls become extinct. By tying itself so closely to its habitat, it becomes superbly adjusted—but at the expense of the evolutionary safety that comes from being able to adapt to many habitats (many species of host).

The dilemma: whether to adapt to a precise niche, changing over time to become ever more perfectly adjusted to that niche, or to use a more generalized approach—adapt to many niches. Which is more likely to accomplish survival of the species over time?*

Schistosomes have "chosen" precision. Consequently, although the cercaria detects a shadow and enters it, if the host is not a seagull it does not find the metabolic signals that mark the pathway to the liver, thence the lower mesenteric blood vessels. It becomes lost. Further, not having adapted precisely to the foreign host (in this case, humans), it sends out antigenic signals.

---

*Both mechanisms have proven to be successful.

PLATE 15

A. HERRING GULL, *Larus argentatus*. To 26 in.; adult has gray back, white underside, head; black tail; orange stripe on yellow beak. A few from the hordes living on mountain of garbage were collected for class study; survey of gull parasites.

B. MUD SNAIL, *Ilyanassa obsoleta*. To 1 inch; black with minute squares on shell. Usually covered with mud and mucus. Has olfactory organ at base of elephant trunk-like siphon; able to smell carrion from afar; when moving, produces a trail of mucus that is sensed by other mud snails.

Host of five or more kinds of larval parasites. Sometimes more than 20 percent of snails can be infected. Sporocysts and redias sometimes extremely dense in liver, yet snails live on. One unusual fluke species in this snail, *Zoogonus rubellus*, has a cercariaeum (no tail) that crawls to marine worms and parasitizes them; reaches adulthood in flounders; another has spiny collar.

C. MUD SNAILS EATING DEAD FISH. Hundreds of snails often accumulate on rotting animal, usually piled on one another.

D. STUDENT DISSECTS OUT SCHISTOSOME, *Austrobilharzia variglandis*, from lower abdominal vein of seagull.

E. SWIMMER'S ITCH. Caused by *Austrobilharzia*, a schistosome fluke parasite of birds. Cercariae penetrate human skin by mistake, provoking immune response of blisters, red pimples, and urticarial wheals (raised red welts).

A.

B.

E.

C.

D.

Herbie had been attacked by these schistosome cercariae in previous play with the snails. There were no symptoms. But their alien emanations had stimulated his immune system to be ready for future attacks. Subsequent attacks by these cercariae triggered the sensitized immune system, and the intruders were attacked by his body's defenses. First, the invading larvae couldn't find the signals for their blood-borne migration. They were lost. Then they were fiercely bombarded by antibodies and ravenous white blood cells as they wandered around under the skin until they succumbed. Their inflamed, itchy, tortuous trail could be traced until their death, demarked by its abrupt end a few inches from the site of penetration.

The perfection of host specificity had protected Herbie. His abortive schistosomiasis was no more dangerous than a case of poison ivy. He quickly recovered from his case of swimmer's (or clammer's) itch.

16

A BETTER MOUSETRAP

A MAN NAMED SHERMAN invented a better mousetrap. It consists of a flat aluminum sandwich that springs open when squeezed, turning it into an eight-inch-long rectangular dead-end tunnel. One end is hinged and attached to a seesaw trigger on which is smeared a dab of peanut butter. The other end is closed. A small rodent smells the peanut butter, enters the trap, and trips the trigger/treadle. The door snaps closed. The mouse is trapped!

I became adept at setting the traps. Removing one from a holster on my waist, I squeezed it. Pop! It became a dully glowing metallic rectangle. I smeared peanut butter on the trigger and placed the trap under a bush, all in a few smooth movements, making it possible to set a mile-long trapline in a couple of hours.

A day later I returned and checked the front door of each trap. If it was closed, something mysterious had sprung it. Was it the rice rat, *Oryzomes palustris*, a white-footed mouse, *Peromyscus maniculatus*, or another of the extraordinarily abundant rodents that abound in fields and forests? This survey of rodent parasites was arduous, requiring marching through swamp and forest for miles each day, rain or shine. Mice and rats have such a high metabolic rate that being confined for twelve hours deprives them of food and warmth and they invariably die. It was necessary to collect them as soon after death as possible, lest the parasites in their intestines suffer the same fate and disintegrate.

One day I heard an uncharacteristic scrabbling in the trap. I opened the door, and a small ground-feeding bird fluttered in the confines of the trap. It had been lured into the trap by that universal elixir of youth, peanut butter.

Back at the lab, I examined the contents of its gut. In the rectal region I saw a digenetic trematode that I recognized as the famous *Leucochloridium macrostomum*, the big-mouthed fluke (macro = big, stomum = mouth).

THE INVASION OF THE FLUKES

Flukes (digenetic trematodes) probably evolved from the same or similar ancestor as the tapeworms. Hence they faced the same seemingly insurmountable crisis in their evolution when their hosts adapted to a terrestrial existence.

The ancestral fluke had to be a parasite of aquatic snails. Why else, to this day, would virtually all digenetic trematodes (*digenetic* refers to a two-step reproductive cycle, each involving a separate host) incorporate snails as the first step of their larval development? But this affection for water-dwelling snails was the undoing of the early flukes. Their larvae were stuck inside aquatic snails. You would think that they would be satisfied with the richly nourishing foie gras of the aquatic snail liver. But no, there were vast numbers of tasty potential hosts on land. Their evolutionary quest would need the participation of a second intermediate host to reach us and our vertebrate brethren.

Many trematode species invented solutions to win the difficult battle. Some were bizarre because the problem was vast, requiring such evolutionary elasticity and inventiveness as to make the transition to the land of free-living species seem trivial. It was necessary for the parasitic trematode to:

1. Penetrate terrestrial snails and eventually use these land snails for the first stages of larval development (first intermediate host).

2. Use the larvae inside the snails as a springboard to infect another animal, the second intermediate host.
3. Use the second intermediate host for a second period of larval development to produce the infective larva.
4. Transfer this larva to the final, ultimately suitable host *without an aquatic connection.*

SPECTACULAR SOLUTION

The aquatic snail stops at the water's edge. Like the tapeworm, a variety of evolutionary solutions was used to bridge the aquatic-terrestrial gap. How does the parasite enter a tempting terrestrial target? Among the strangest methods is that evolved by the digenetic trematode, *Leucochloridium.*

A bird defecates on leaves. Land snails, species of *Succinea*, eat the egg-laden feces together with the leaf, like caviar on crackers.

The eggs hatch into two fecund larval stages in the snail's gut. No swimming needed. But emergence from the water has not yet been achieved. The standard swimming stage is still stuck inside the land snail. How to enter the final host to at last complete the cycle *without water as a transfer medium?* Inside the snail the parasite's cells appear to divide in an uncontrolled manner like cancer cells. Unlike cancer, no tumor is produced. Instead, the resulting growth becomes a sausage, filled with other sausages, the mother and daughter sporocysts.

The daughter sporocysts usually give birth to hundreds of infective swimmers called cercariae. Traditionally they would enter a minnow or frog and encyst, waiting to be eaten. But this snail is on land. Its burden of cercariae cannot be shed into water. Nor can they be deposited on land lest they dry up and die.

Leucochloridium has ingeniously solved the problem. The second sausage stage (daughter sporocyst) becomes filled with infective cercariae as usual. Then it morphs into a monstrous, cancerlike intruder, filling the host's liver and body cavity like the parasitic beast in the movie *Alien.* A branch randomly grows toward the head and penetrates one of the snail's antennae. There it fills with cercariae

PLATE 16

A. BIG-MOUTHED FLUKE, *Leucochloridium macrostomum*. To 3/8 inch; large oral sucker contains mouth; black irregular bands along sides are yolk glands; central donut is ventral sucker; below it are ovary (*left*) and two testes (*right*); uterus is tortuous tube opening at posterior (projection outside body is artifact). In rectum of birds.

B. BIG-MOUTHED FLUKE DAUGHTER SPOROCYST. Cancer-like tubes grow throughout body of snail; randomly spread until they penetrate antennae and siphon.

C. COMPARISON BETWEEN INFECTED AND UNINFECTED SNAILS. Antennae and siphon of infected snail (*right*) swollen; skin stretched to transparency showing pulsating orange and green-striped sporocyst containing cercariae. Birds swallow cercariae; they become adults that inhabit bird rectum.

D. ANT ON BLADE OF GRASS ate cercariae of cattle liver fluke, *Dicrocoelium dendriticum*, buried in balls of snail slime; cercariae burrow into ant's brain, causing ant to climb on grass, attach with jaws; ant eaten by cows, sheep.

E. ANT, *Formica* spp., eats balls of snail slime (shown) containing cercariae

F. SHERMAN TRAP. 10 by 3 by 3 inches, aluminum. In right hand drawing it is closed, 3/4 inch thick. Pressure on trigger snaps trap open, revealing treadle, door; when rodent touches bait on treadle, door snaps shut.

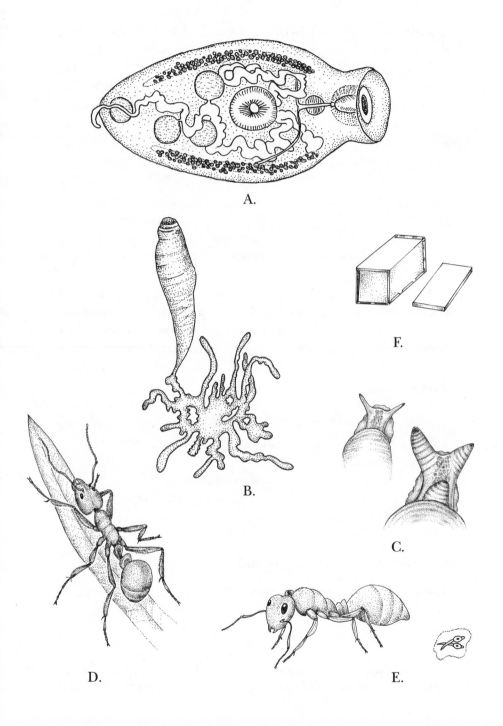

A.

B.

C.

D.

E.

F.

and expands massively, stretching the skin of the antenna until it becomes almost transparent. Amazingly, in this species the normally nondescript sporocyst is gorgeous, striped with bright orange and green bands, brightly visible through the walls of the swollen snail antenna, the skin now stretched to transparency. It begins to pulsate! A small bird sees this bright pulsating morsel. It eats it. Presto! The cycle is complete.* In a few weeks the bird's cloaca (rectal region) contains many mature, egg-laying, big-mouthed flukes.

CRAZY ANTS

The ultimate terrestrial transfer harnesses two intermediate hosts, an ant and a snail, in a desperate Byzantine partnership to reach the blood-rich liver of a cow.

The tiny cattle liver fluke, *Dicrocoelium dentriticum*, lays eggs that contain a ciliated larva, the miracidium, that in aquatic species would normally swim to a snail, infect it, and start the cycle. In this case, it remains in the egg when it is released in cow flop. Terrestrial snails, species of *Clionella*, gather at the fecal feast and eat the eggs and their miracidias, beginning the cycle.

Eventually the miracidias become mother and daughter sporocysts in the snail's liver. The daughters produce vast numbers of cercariae. These migrate to the snail's "lung," irritating it. The irritated snail responds. It covers up to a hundred cercariae at a time with slime. The snail crawls along, leaving its characteristic trail of slime.

Buried in the slime are balls of cercariae, which seem to be particularly delectable to ants.

HUNGRY ANTS AND SLIMY BALLS

The ant, *Formica*, eats the slime balls, freeing many cercariae into its intestine. These usually burrow through the gut walls and encyst in

---

*Should a bird not peck up the sporocyst-laden antenna, it stretches and stretches until it bursts, showering encysted cercariae over nearby tasty leaves and bushes.

the ant's body cavity, becoming infective metacercariae. But a few wander to the ant's brain. It begins to act strangely.

Instead of joining its brethren and returning to the colony, the infected ant climbs to the top of grass blades and locks its jaws powerfully on the grass tip. So strong is this grasp that if you pull on the abdomen, the body will break before the jaws let go. The jaws stay locked all night. At sunrise when the ant is warmed, it lets go and crawls down the stalk to the cool shade, only to return to the tip at dusk. When a hungry cow eats early breakfast or late dinner, it is vulnerable. The cow (or, rarely, human) eats the infected ant with the grass. The encysted metacercariae are freed in the cow's gut and end up in its liver. The cycle is closed without water being involved.

The bridge over troubling water is complete.

17

SCANDALS AND GHOSTS

THE MUTED SCREAM STOPPED, severed by the muffled murmur of a hundred excited voices. Where was the pandemonium coming from? Suddenly the ornate double doors of the auditorium next to my lab burst open and a crowd of well-coiffed, elegant-looking, matronly women emerged. Faces flushed and clearly upset, they were literally carrying one of their own, her arm held over her eyes as if mocking the heroine of a silent movie. What tragedy had befallen her? What had led to the emotional outburst that filtered through the walls of my laboratory? Later on, I found out that my fate hinged on the very serious origin of this meeting of a hundred obviously well-to-do women.

The source of the crisis was the perceived deception perpetrated on them by the flamboyant director of the laboratory. In his desperate search for funding, he had resorted to a time-honored method used by small research organizations struggling for support. He approached the private sector—in this case, an organization of hundreds of women whose role was to raise money for a noble cause, the perpetuation of the medical foundation that he represented. Every year the women would meet to organize a charity ball to raise money for the institute's research on a promising mechanism for controlling a form of cancer.

Funding agencies don't like to support the investigation of unproven therapies (which would seem to be counterintuitive), hence the need for large sums of nongovernment money. The charitable instincts of the women were focused on the promise of a cure for cancer. In speeches to his all-female audience, the director made statements like, "We don't want you to smoke because we love you." Evidently some of the women took the word "love" personally and became involved on an emotional level. I found the director to be pleasant-looking, but no Adonis. But who can understand these attractions?

A week before the meeting, a newspaper photograph had shown the director boarding a plane on his way to a medical conference. Next to him was the administrator of the lab, a chubby woman in her late forties, the femme fatale of the drama. (Who can understand these attractions?) Everyone believed that they were running off together to the conference and that charitable funds were being used to foster the torrid affair.* The women, distraught and resentful, had their emotional meeting and decided to cease their involvement. Without funds from the charity ball, the institute languished and disappeared. So much for the support of cancer research.

I was incidentally a victim of the institute's demise. My borrowed lab disappeared when the institute disappeared.

SLIPPERY SURFACE SYMBIONTS

My research at the institute wasn't on cancer. I was interested in an obscure group of parasitic worms that usually live on the outside of fishes, on skin, gills, even eyes. This group of external flukes (monogenetic trematodes) may have coevolved from the ancestral precursor of the internal flukes that wreak so much havoc on humans—or not.

Monogenetic trematodes don't require intermediate hosts. The more familiar digenetic trematodes usually require two. Monogenetic trematodes live on the *outside* of fishes, and their digenetic relatives live on the *inside* of most vertebrates (including humans). Could both types of worms have arisen from an ancient common ancestor and

---

*There was no evidence, to my knowledge, that this was indeed a tryst or that the funds used were institutional and not private. Besides, what is so wrong with a director traveling with an administrator? But rumor had its evil effect.

chosen divergent life styles? Or had they evolved anew, unrelated to each other?

My work, perhaps, could contribute to the solution to this question.

Many monogeneans are thought to eat the host's blood. Others are believed to eat skin cells. Some, preferring the gill cavity, eat gill epithelia. These are the fortunate, for they are protected from the rigors of living on the skin as the fish hurtles through the water.

Slimy skin seals the surface of the fish, preventing invaders like the cottony parasitic mold, *Saprolegnia,* from sucking sustenance from the fish with rootlike tendrils. The slime is mucus; the fish is literally encased in this slippery substance produced by an underlayer of epithelial cells. Do these nutritious epithelial cells sustain the worms, or is it the mucus—or both? Unlike their internal brethren, these ectoparasitic worms need not figure out how to enter the body or pass through the fiery pit of the stomach. They can survive on the ever-present skin epithelium. Mucus is made of carbohydrates, a fitting addition to the worm's meal. The epithelium is the steak and the mucus, the potatoes—meat and carbs, an all-American meal.

But there is another problem that must be overcome before these surface feeders can find parasite paradise. And paradise it is, because the skin constantly renews itself by sloughing off its outer layer, so that epithelial cells eaten by the worm are fresh. Little harm to the host— an unfettered future for the worms. What's the catch? There has got to be a catch. Life in slime may be paradise, but slime is slippery—and the fish swims fast, too fast for safety.

Evolution provides. The worms uniformly possess a posterior structure called an opisthaptor. This comparatively huge, flat fleshy plate is armed with a variety of fearsome attachment structures—hooks, anchors, suction cups, clamps—designed to keep the worm on its little universe. No matter how fast the fish swims, and a tuna can swim thirty miles per hour, the opisthaptor grips the surface tenaciously.

KILLING THE UNKILLABLE

The monogenetic trematode I was studying, *Gyrodactylus,* has both huge curved anchors and tiny hooks decorating its huge flabby plate-like opisthaptor. It lives in considerable numbers on the surface and in the gill cavities of the common killie or mummichog, *Fundulus heteroclitus*, a minnow so resistant to environmental vicissitudes that it virtually can't be killed—except by me.

Early on I realized that research is a monumental task, often producing no useable data, wasting months of effort. Consequently, I resolved to seek every advantage. In my case that meant finding research organisms that are so sturdy as to be able to survive anywhere under the most impossible conditions. Cockroaches fill the bill. So do their aquatic equivalents, killies. These small fishes can be found in mosquito ditches in marshes and in shallow water anywhere the water is even slightly salty. When summer comes, they can be found in tide pools that sometimes become isolated from the sea. Temperatures reach 100 degrees Fahrenheit as the sun beats down on the shallow pool. The water begins to evaporate and becomes very salty. Oxygen levels decline drastically. Still the killies survive. Fishermen use them for bait, keeping them alive in moist seaweed for a day.

That's the organism for me, I thought. They are called killies to defy the gods because they can't be killed. Having little research money and no women's organization to raise money for my studies of obscure parasites of fishes, I found inexpensive containers to hold my research organisms—kiddie pools. The formula: take some plastic pools and fill them with aged salt water. Add fish.

The next day I returned to the lab, only to find killies strewn all over the floor, stone dead. They had leaped out of the water, committing suicide. Those in the pools were swimming listlessly, obviously in the last stages of life. *I had killed the unkillable.*

Each time I stocked a kiddie pool with killies, they were dead within forty-eight hours. If I can't keep killies alive, what hope is there for my research? I was depressed. One day, several months and hundreds of dead killies later, I read a paper on culturing killies. These researchers, too, had little money. They too had used kiddie pools to hold their fishes. It pays to read scientific papers carefully. There, in a small print footnote, was a sentence: "Do not attempt to raise fishes in pools meant for children unless they are conditioned for three months. The plastic polymer contains a substance that leaches from the walls, designed to prevent algal build-up in the water." No wonder my killies died. They were unkillable in nature, but when placed in man-made pools, they dropped like flies. I wondered what effect the algicide had on kiddies.*

After filling and refilling the pools for months to leach out the algicide, I was ready to start.

All this death-dealing activity had occurred before I moved to the institute. I had resolved two practical problems: I was able to culture the killies, and I had become familiar with the life cycle of the parasite, *Gyrodactylus prolongis*.

I was ready to work at the institute with its sophisticated equipment and know-how.

NO PRACTICAL PURPOSE

A scientist does not set out to find a "cure for cancer." He or she chooses to investigate a problem inspired by an interest in a field of study *with no practical intent*. A scientist is supposed to be like a curious child, investigating whatever strikes his or her intellectual fancy.

In these modern times, even an intent to answer a scientific question is directed by Big Brother. If the researcher does not envisage a "practical" purpose, it will be difficult to obtain the increasingly large amounts of funds necessary to acquire sophisticated instrumentation and graduate students.

People like me, who wander through field and feces, are anachronisms. I am interested in the purest of science, the resolution of problems that have no practical consequence. To my mind, science should have no agenda. Nor should a researcher direct his or her investigation toward a particular complex goal, for example, "finding a cure." Science is incremental; resolution of a problem by one person contributes to the resolution of the next person's problem, and so on. Eventually someone sees the big picture and ties together his predecessors' work into a major discovery. She gets a Nobel Prize—on the backs of her predecessors.

---

*There is some consolation in the fact that the fishes take in water all their (short) lives, while children, at worst, swallow a few mouthfuls a year in summer.

Nowadays it is practical to choose a "fashionable" field that has "useful" consequences in order to have a good chance to get funding. Until a year or two ago it was difficult to study stem cells. Funding depended on political interference in a ghastly distortion of the essence of science. If even one person in a lab was performing "prohibited" research, the whole lab, with its many scientists, could have lost government support.

Something similar happened in Russia in the early days of communism. A powerful, politically connected geneticist named Lysenko pushed back genetic research there for generations. This state-sponsored kind of interference goes along with past political positions tinged with the preposterous notion that "intelligent design" is science.

My unfunded project had no obvious practical consequences. The question I was trying to answer was: Is it possible to culture monogenetic trematodes in a dish? That's all. Not to develop a cure for an ailment, not even of the fishes upon which they live. (Although, years later, a species of *Gyrodactylus* decimated salmon farms and a tissue culture technique was needed to find an agent that would prevent the epizootic [epidemic]. My previous pointless-seeming system might have been part of the solution.)

The first step would be to culture killie skin cells; to develop a "line" of skin cells and grow them in a dish. Then to use a layer of these cells as a substratum for the monogenetic trematodes to live on, like a herd of cows eating grass.

We cultured skin epithelial tissue from isolated cells extracted from mature killie skin. In the case of human stem cell culture, cells are taken from an early human embryo, sometimes a microscopic ball of a few cells. These cells are destined to become the tissues of the body, but at this early stage they can become any kind of cell, hence the term "stem cell."

I was trying to culture an already differentiated tissue, skin epithelium. If successful, I planned to place the monogenetic trematodes on a layer of this tissue. *Someone else* might use this technique to find out answers to questions like these:

What harm do the worms visit on their host?

Is the rate of growth of the skin cells sufficient to replace those that are eaten in order to maintain the protective mucus layer?

How many worms can parasitize the fish before its skin can no longer ward off the harmful pathogens lurking nearby in huge numbers?

Does the worm eat cells or mucus, or both? By coloring the skin cells with a harmless stain, it would be possible to trace the movement of these cells passing through the worm's gut.

My first problem: how to obtain healthy worms. The traditional method of removing monogenetic trematodes from their host fishes was developed by fish farmers. The farmer added toxic formaldehyde to the water at concentrations high enough to kill the worms but not the fishes. That technique did not yield the living worms I needed. I would have to develop a new technique. At a loss, I consulted the aforementioned director of the facility, a famous physiologist. Perhaps his point of view would help. He offhandedly said, "Try using antihistamines, they relax muscles."

I purchased over-the-counter antihistamines and added various concentrations to the culture water. Voila! At the bottom of the dish I found dozens of healthy, writhing worms. The muscles operating the hooks on their opisthaptors had become so relaxed as to cause them to fall off the host fishes. I had the worms for step one!

Now how to feed the guys. *Gyrodactylus* eats skin epithelial cells (or the mucus they produce?). That meant I had to produce the cells in flat, skinlike sheets independent of the host fish. There exist well-described techniques for culturing tissues. Surprisingly, some have been developed for killie epithelia.

Using one of these techniques, I cut off small pieces of skin and placed them in an antibiotic solution to kill any surface bacteria. Then I dropped them into a mixture of antibiotics and culture medium and whirled them in a kitchen blender, separating the cells.

A slurry of separated cells was placed in a multichambered dish containing the culture medium. Streptomycin and penicillin were added. (It is interesting to note that no matter how sophisticated the chemical composition of the culture medium, it still requires at least 10 percent bovine serum [plasma from the blood of a calf]. One cannot help but think that there is some mysterious substance in the blood that adds that indefinable "life essence.")

I was ready to "plate out" the skin epithelial cells. A few were removed and dropped into sterile medium in a dish.

After many hours, isolated cells became visible. They coalesced into a sheet of epithelial cells in a day or two. I became adept at maintaining this tissue culture.

RUSSIAN DOLLS

The next step was to obtain the hundreds of worms I needed for the investigation.

I knew that *Gyrodactylus* has a miraculous method of reproduction that would solve my problem. I had stacked the cards in my favor by choosing this species.

In the wild, few of these parasites are found on each fish. When crowded, as in fish farms, the fishes are so tightly packed that they are literally cheek by jowl next to each other. *Gyrodactylus* cannot infect fish separated by even a small distance—contact is necessary. In an extraordinary variation from the reproductive pattern of other monogenetic trematodes, they give birth to live young, forgoing the usual motile infective larval stage.

I crowded lightly infected fishes for a day or two. Then I placed each previously crowded killie in its own dish. When I added antihistamines a few days later, a miracle had occurred. Vast numbers of worms carpeted the bottom of each dish. How could so many appear so soon?

When examined under the microscope, *Gyrodactylus* looks like a ghost. It is easy to identify each internal organ through the translucent body. The uterus takes up a large portion of the internal cavity. Inside the uterus is the soon-to-be-born infant. *But inside the infant's uterus is its offspring, and inside the offspring's offspring is one final offspring.* There were four juveniles inside each adult worm packed like Russian dolls. No wonder they had so densely infected the killies so rapidly.

Ordinary animal development originates with the fertilized egg (zygote). It divides to produce a two-celled stage, each identical cell (blastomere) attached to the other, each destined to give rise to future organs. The whole animal is preprogrammed in these two cells. In humans the cells sometimes do not stick to one another. These go through normal development, each becoming an identical twin. But

PLATE 17

A. KILLIE, MUMMICHOG, *Fundulus heteroclitus*. To 6 inches; male green with brown stripes, white spots; mating colors of dorsal, ventral fins have black-and-white spots; female pale green, larger than male. Small mouth; omnivorous.

B. LIVE-BEARING MONOGENEAN, *Gyrodactylus prolongis*. To 1/16 inch; robust anchors. Scattered anterior cephalic glands produce mucus for attachment at front. "Barrel" is mouth and pharynx; gray, Y-shaped organ is intestine (no anus); within forks of Y is uterus showing an *embryo* that contains three other embryos, each inside one another. Four generations of worms in uterus of mother; all have both male and female gonads. All act as females and produce offspring without being fertilized, but in third or fourth generation some act as males for sexual reproduction (my interpretation—process not well understood).

Below uterus are white paired testes and single dark ovary. Opisthaptor at rear used to anchor worm on killie's slimy skin, gills; armed with sixteen peripheral hooks and two anchors.

It is now known that worm feeds by everting pharynx through mouth, secreting proteolytic enzymes, digesting both mucus and skin, sucking resulting fluid into gut.

C. PARADOXICAL DOUBLE DIPLOZOON, *Diplozoon paradoxum*. To ¼ inch; on fish gills; feeds on blood. A larva cannot survive unless it meets another; they fuse; each grows an opisthaptor. Male and female genital ducts nearly fuse; after cross-fertilization each produces eggs that remain on host's gills.

"Grapes" in body are yolk glands; tubes are genital ducts; dark circles are ovaries; groups of cells anterior to them are testes; opisthaptors have eight clamps, each with a small hook. Worms live in permanent copula.

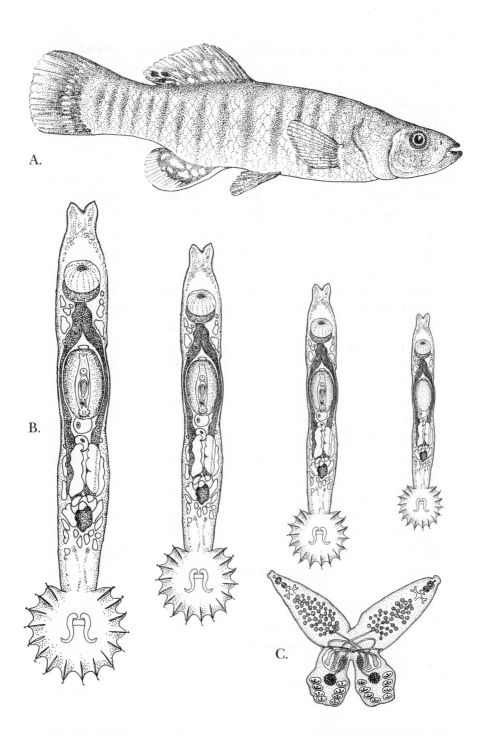

A.

B.

C.

in *Gyrodactylus* the usual disciplined pattern of blastomere development does not occur. Using a unique process of cloning similar to
the polyembryony of digenetic trematodes, identical quadruplets are
produced, each larger and more well developed than the other, *each
inside the other*. On the day of its birth, each quadruplet can potentially
give birth to three other offspring, and within a few days all four are
chewing away on the surface of the fish alongside the mother.

Birth seems to have little effect on the moms. There is no sign of
damage when the daughter bursts free. Within twenty-four hours,
daughter and mother are feeding side-by-side. Then, in an extraordinary evolutionary triumph, inside a placenta-like structure in
daughter one lies daughter two. Nourished by this "placenta" lining
the uterus, daughter two is ready for birth. Inside her uterus lie sisters
three and four. Naturally all clones of the mother are identical to her,
but all are hermaphrodites, capable of sexual reproduction. For survival of the species it is necessary for a worm with a different genome
to be part of the population on the killie. When the host fish rubbed
against another, several worms were transferred, including one with
a different genome, to act as a male. The laws of evolution must be
obeyed even in this microcosm on the flanks of a fish.

This explains how several worms transferred to a previously uninfected killie can, within a few days, produce a horde of miniature
monsters, each parasite inch-worming its way over the surface, slurp-
ing slime and skin cells, leaving the fish vulnerable to secondary infection. So weakened, it can die. It is no wonder that trout, salmon,
and goldfish hatcheries are plagued by *Gyrodactylus* epidemics. In the
twenty-five years since the introduction of *Gyrodactylus salaris* from the
Baltic Sea, Norwegian salmon farmers lost 500 million euros from infestations of salmon crowded in rearing cages.

In a few days I had enough worms to perform the next steps—how
to make the worms bacteria-free so as not to contaminate the sterile
fluid of the tissue culture, and the final step—to actually introduce
the worms into the culture dish.

Then the institute closed down.

18

SPINY-HEADED MONSTERS

THE BIOLOGY BUILDING may have dated from the founding of the university in 1884. Halls were narrow warrens of mustiness and decay; shiny black floors were layered with scraps of linoleum. High walls, painted institutional green, had turned grayish with dust. In this ancient environment my colleagues were attempting to perform cutting-edge science. An electron microscope was quartered downstairs.

The bathroom on our floor was a green cave illuminated by a few dim bulbs from the ceiling high above. The light seemed to be sucked in by the now dull, formerly shiny walls of the stalls. The urinal was a simple trough with a thin stream of water running perpetually along its bottom. Above the white stone sink was a pewter-colored rusty soap dish with a sticky bar of well-used white soap. It had deep, flat grooves chiseled into one edge. Suspicious, I brought the soap back to the lab. "What made these grooves?" "Rats" was the immediate answer.

BATHROOM SAFARI

My self-imposed research project was a survey of internal parasites (endoparasites) found in rodents in the southeastern United States. "At least it will get me outside," I thought. The project entailed capturing rodents—mice, rats, voles, and related animals—in traps, terminally anesthetizing them, and performing a postmortem on their internal organs. This method is traditionally used to discover new species or extensions of previously described ranges.

The aforementioned presumptive rat was a rodent. It was fair game.

Wintry rain was falling; dark, foreboding skies made me shiver. I really didn't want to trudge through a muddy marsh to follow my trapline. Why not do it the easy way and trap the rat in the bathroom? The next morning, I entered the rat's den—the gloomy bathroom— and found the trap sprung, its peanut butter bait evidently a better choice of food than Ivory soap. Crammed into the live trap was a fat, ragged old rat. Its long, scaly, gray-pink tail hardly fit. I have a perverse fondness for vermin and felt a twinge of regret as I performed the postmortem.

Lungs—negative—check; pleural cavity—negative—check; abdominal cavity—negative—check; small intestines—bingo! Attached to the inner wall of the upper small intestine (duodenum) was what appeared to be a tiny string of pearls. Had the rat eaten a child's doll, its undigestible jewelry now glowing from the rat's gut? Glistening with white, translucent iridescence was a miniature succession of regular swellings alternating with corresponding constrictions. Wait, it moved! I looked again. These twitches didn't match the normal peristaltic undulations of the intestines. This white thing was alive. It was some kind of worm. I thought I knew a lot about worms, but this was new. Using forceps, I tried gently to remove it. It wouldn't budge from its attachment to the duodenal wall. My mentor looked over my shoulder, intrigued by this unusual discovery. It was huge, about two inches long and fat enough to stuff the gut of the rat. I continued to tug. "Stop!" he shouted. "You'll break it!" My ears rang with his warning. "The only way you can free it is to cut out a chunk of intestine around its attachment and keep it in distilled water overnight."

What is this arcane technique? I thought. I had never heard of releasing worms by putting them in distilled water. The vaguely similar tapeworms attach to upper intestinal walls like this mysterious animal, but weakly, often to lie free in the gut, traumatized by their violent exposure to the outer world. This animal was different. It was literally stuck to the wall.

I followed the wise counsel and the next morning returned. On the bottom of the dish was the pearly worm. Scarcely visible at one end was an elongated pimple. Under magnification the pimple became a hideous "head" composed of row upon row of sharp, recurved spines that had been buried in the intestinal wall. This monstrous means of attachment was more effective than the relatively benign suckers of a tapeworm or a fluke. This had to be another phylum. Dredging up an obscure memory, I realized that this was an acanthocephalan, a thorny-headed worm (acanth = spine, thorn; cephala = head).

Its "head" had been deeply embedded in the intestinal wall. It would have been impossible to remove intact except by the obscure distilled-water method, used only for acanthocephalans, unknown to any but specialists. The theory is to force it to relinquish its hold by hydraulic pressure. By placing it in distilled water, its cells imbibe the water and it swells, pressing it against the intestinal wall with inexorable pressure. Inevitably, it retracts its spines and literally pops off the gut wall, ending up supine on the bottom of the dish.

Acanthocephalans live in the gut in the manner of tapeworms, but that is where the resemblance ends. The body surface is riddled with pores rather than the fuzzy extensions, microvilli, of tapeworms. The host digests its food, but the acanthocephalan further breaks up the molecules at the surface, and only the tiniest fundamental nutrient molecules pass into the body. This is fortunate, for there is no mouth or gut. The pores open into tubules that pass through the skin to fork into blind passages in a "felt layer." Wandering cells, similar to human white blood cells, randomly carry the now thoroughly digested food through the fluid-filled body. These cells are not so unusual. They are found throughout the animal kingdom, starting with the most primitive animals, sponges. Some human white blood cells, collectively called amebocytes, can pass through the walls of blood vessels like ghosts—a phenomenon called diapedesis—leaving no hole to mark their passage.

A tapeworm doesn't have a lot of structure, but acanthocephalans do them one better. With reduced muscles and no appreciable digestive, excretory, circulatory, or nervous systems, the acanthocephalan body is just an elongated sac of gonads. There is usually a thread, the suspensory ligament, plus two invaginations of the body wall and a sac that is used to contain the withdrawn "head," which is really just a spine-covered proboscis.

A CONVENTIONAL SEX LIFE

Acanthocephalans have an unusual sex life compared to hermaphroditic tapeworms and flukes. There are males and females. The male has a penis; the female, a vagina. The male organ of the hermaphroditic tapeworm is called a cirrus. It extends by evaginating (turning inside out like a child's party favor). Roundworms have both sexes, but they use copulatory spicules that are "horny" spikes inserted into the female's vagina to act as a bridge for the ameboid sperms to wander across. One is tempted to assign an evolutionary significance to the advent of the acanthocephalan's penis, but human appreciation of this organ notwithstanding, it is not symbolic of a superior evolutionary position of this phylum.

Further "romancing the phylum" reveals that sex is performed "the regular way." The male acanthocephalan uses its penis to insert sperms into her vagina. However, along with his sperms his semen contains an unusual "copulatory cap" made of protein that prevents further copulation, forcing the female "to be true to her partner," passing on his genes and no others. The cap dissolves before the female releases the eggs.

The microscope reveals that the body of an adult female is distended by thousands of fertilized eggs that she releases daily into the intestines of the definitive host. These then pass out of the body in the feces, to be eaten by an intermediate host like insects or minnows for further development. But acanthocephalan wizardry is further exhibited when the rat eats too many infected roaches. Their daily bug meals inundate the gut with juveniles seeking a nice spot to spend the next few years soaked in digested food. The new crowd threatens to overwhelm the indigenous population in the upper intestine. In the face of this threat, the acanthocephalans regulate their own population, not by some chemical means but by the fact that there are only so many desirable spots in the preferred region. Subsequent juveniles looking for a place to attach cannot find one and are swept out in the feces.

DO NOT LET YOUR CHILD EAT COCKROACHES

Acanthocephalans probably originated as fish parasites, for adults are found in many species of fish and fish-eating birds. Only a few inhabit terrestrial hosts. Fewer are found in mammals. The impressively named pig acanthocephalan, *Macranthorhynchus hirudinaceous*, can be a foot long. The unusual relationship probably evolved from the pig's less-than-scrupulous food choices, like eating cockroaches. Same for the rat acanthocephalan, *Moniliformis moniliformis*. (It seems that acanthocephalan taxonomists are supernerds. Not caring about the masses, they assign giant names to tiny worms.)

A cockroach, to no one's surprise, eats anything, including rat feces. Once eaten, eggs in the feces hatch into larvae that penetrate the gut wall and rest in the body cavity. A rat then eats the cockroach. Once inside the rat's gut, the acanthocephalan becomes sexually mature and releases huge numbers of eggs that are mixed in with the feces. All this expenditure of energy is needed to produce enough eggs to increase the odds of a chance encounter with one cockroach that eats an egg in order to complete the acanthocephalan life cycle. How risky is that?

Occasionally a child eats an infected roach. If the child is malnourished, it exhibits symptoms like a distended abdomen, diarrhea, vomiting, "foamy feces" (poetic, but what does it mean?). The mother of a healthy fifteen-month-old baby boy in Florida found white worms in his diapers. Upset, she reported the condition. The child, showing no additional symptoms, was treated with the standard medication, mebendazole. An additional twelve adult *Moniliformis* were passed, but another medication and two more months were needed for full recovery.

BLOWING WIND AT THE ASS OF A COCKROACH

Everyone hates a roach, even me. I have spent my career culturing and living with them in close quarters in my lab, but I still hate them. Others who have accepted the disgusting burden of working with roaches have discovered that cockroaches infected by *Moniliformis* respond to the breath of the rat predator. Do rats characteristically have bad breath? Very likely, but further study revealed that it is the very act of breathing that causes the roach's response. There is logic behind the response. The anticipatory heavy breathing of the rat before it pounces provokes a millisecond-level escape response. This normally can save the roach. But sometimes the behavior of the roach seems to invite its destruction and it ends up as a juicy, proteinaceous meal for the rat. Why do some roaches survive the attack and some seemingly invite death? Are they deliberately committing suicide?

THE SUICIDE HYPOTHESIS

A creative study involves blowing wind at the ass-end of cockroaches to test the "suicide hypothesis." The posterior of insects has a pair of structures that look like the curved horns of a cow, leading the uninitiated to confuse front with back. These are called "cercae," and they function primarily to detect disturbances (vibrations) in the air, acting vaguely like a human ear. One roach survives by a hair's "breath," detecting vibrations with its sensitive cercae, while another sluggishly succumbs.

If you attempt to stomp a roach to death, its cercae will detect the rush of air descending from your downward-moving shoe, and the roach will evade your attempt. Survival is dependent on a split-second response. But you succeed some of the time, leaving a disgusting stain on the rug.

The study showed that the presence of infective larval acanthocephalans in the body cavity of the roach slows its behavior, allowing the predator (the rat, bird, cat, or child) to capture and swallow it. Some infected roaches seem to seek out a light-colored surface, standing starkly visible, as if paralyzed. The roach is literally committing suicide.

The apparently suicidal "intent" of the parasitized roach causes it to:

1. spend more time on horizontal surfaces.
2. seek out and stand immobilized on light-colored surfaces that make it more visible to predators.
3. move less and respond more slowly.*

---

*It takes less energy to remain horizontal. To achieve verticality would tax an infected, weakened roach. To move less frequently and more slowly would also conserve energy. But to seek out light areas is not explainable by the expenditure of energy hypothesis.

Is this really an attempt by the parasitized roach to commit suicide, thus ending its tortured existence?

Or maybe roaches *benefit* from suicide by reducing the number of susceptible members in the population, thus evolving in the direction of reducing genetic vulnerability to the parasite.* After all, is it not likely that the death of a distinct cohort of a population will tend to reduce the tendency of "bad" (vulnerability) genes, conversely increasing the proportion of "good" (survival) genes?

CONTROLLING THE HOST

The presence of *Moniliformis* in its body cavity makes the roach more vulnerable to the rat in which the parasite will complete its life cycle. Whether or not this phenomenon benefits the parasite or the host or both is up for grabs. But it has been shown that unusual host behavior is beneficial to a species of wasp that lays its eggs inside an aphid. The egg hatches and the larva obtains its sustenance by burrowing through the body of the aphid. Infected aphids become drab and hide, lessening the chance that another wasp will inject its egg into the aphid. Competition between two or more wasp larvae burrowing through the insides of the aphid will endanger the original larva. Thus it is advantageous to eliminate competition with another larva. Is the infected aphid changing its behavior to help its parasite survive?

Infected worker bees tend to be found outside the nest more often than healthy workers. This may prevent them from infecting others in the hive. Thus the unusual behavior of the host species may work both ways, toward survival of the parasite *and* survival of the host.

Here are other examples described in a "charming" newspaper article.

---

*Evolution is not directional. That is, a giraffe does not evolve to eat leaves at the top of trees, thus occupying a less competitive niche. But evolution is directional in the sense that randomly acquired genes that enhance survival accumulate until a species is better adapted to its environment.

PLATE 18

A. BROWN, NORWAY RAT, *Rattus norvegicus.* To 15 inches, nose to tail tip; coarse gray, brownish fur; transmits bubonic plague via oriental rat flea; transmits rat bite fever via bacteria in saliva. Albino important research animal; most common U.S. rat.

B. AMERICAN COCKROACH, *Periplaneta americana.* To 1¼ inches; brown, yellow margins on front; can fly; common in South, anywhere warm. Eats acanthocephalan eggs in rat feces; eggs hatch in roach gut, releasing larvae that encyst in body cavity. When roaches are eaten by rat, larvae excyst, become adult in rat gut.

C. BEADED RAT ACANTHOCEPHALAN, *Moniliformis moniliformis, M. dubius.*

Males to 5 ½ inches, females huge, to 10 ½ inches, fill rat gut; white, beaded; twelve rows of spines on proboscis that can retract using diagonal muscles on sides of sheath (shown); sexes separate; no circulatory, excretory, digestive systems; female's body filled with thousands of eggs released every day via birth pore.

D. ROACH ON WHITE BACKGROUND. Infected with acanthocephalans. Seeks light surface, moves and responds more slowly than uninfected roaches. Suicide?

E. HORSE-HAIR (NEMATOMORPH) WORM, *Spinochordodes tellinii.* To 6 inches; eggs eaten by grasshopper develop into long, thin, "horse hair." Worm fills body cavity, bursts from insect in water, becomes adult, mates in water. Infected grasshoppers often jump into water. Ancients thought living worms arose spontaneously from (nonliving) hairs shed by horses into water. Reality is just as fantastic.

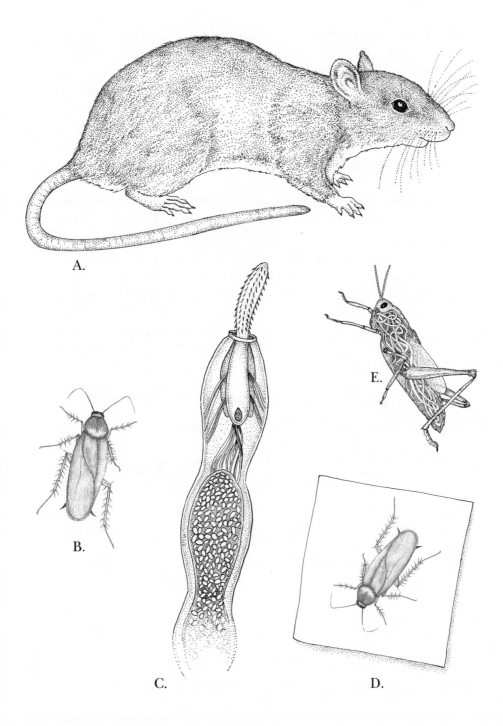

A.

B.

C.

D.

E.

A . . . thorny headed worm . . . spends its childhood growing in the belly of a pill bug. . . . On reaching parasitic puberty, the worm requires a very different environment: an avian digestive tract. The worm wants to be eaten by a bird. Fine: many birds will eat pill bugs if given the chance. But the worm has a tight schedule to keep, and can't afford to wait around . . . the impatient parasite induces a precise sort of madness in its host. It prompts the poor creature to defy all known pill bug precautions, to venture forth in pure daylight and crawl onto a light-colored surface where the dark pill bug stands out so starkly it might as well flash a sign saying, Snack Time! As a result infected pill bugs end up being eaten by birds at a far higher rate than their unparasitized peers. [Similarly] . . . the . . . horsehair worm [phylum nematomorpha] thrives as a larval youth inside a . . . grasshopper, but must be delivered to a stream or lake to breed. Healthy . . . grasshoppers are devoutly terrestrial and do not go swimming. A parasitized insect, however finds itself with the odd compulsion to head for the nearest body of water and jump in,* at which point the ripened writhing worm bursts its host apart like a comic book villain . . . [and seeks a mate].[19]

You gotta believe that parasites can control the behavior of the host.

---

* A recent paper points out that the grasshopper (cricket, in this case) does not seek water but exhibits erratic behavior that accidentally brings it near water. At this point the aberrant behavior kicks in and the grasshopper jumps into the water.

19

BLOODSUCKING BEASTS

THE SICKENING, SULFUROUS STENCH that accompanied each muddy step waned as the patrol emerged from the mangrove swamp. Twisted aerial roots hanging from the densely packed trees no longer grasped at the infantryman's fatigues, now blackened with stinking mud to his groin.

Jungle replaced swamp, and the sweet, fecund aroma of rotting organic matter supplanted the sharp, sea-soaked air of the swamp. It had rained that day, the deafening tropical downpour that was part of daily life in Vietnam. Every leaf dripped; breathing the thick, humid air was exhausting. He peered through the rising mist, searching for an enemy enshrouded in its whitish vagueness. The patrol wended its way through the lush undergrowth. Drops of rainwater spilled from the dark green canopy above, running down his neck and soaking his flak jacket. He blinked, his vision blurry with runoff from his brow. In the perennial twilight the undergrowth strained toward every ray of light with thick, glistening wet leaves.

The squad leader signaled a rest stop. Sweating soldiers dropped to the mud, not caring as the leaf rot pillowed them in its rich brown embrace. An infantryman idly glanced at his buddy lying against a wet tree trunk. A brownish, branch-colored, finger-long THING protruded from the guy's neck. "Another one," the soldier spat, grimacing with disgust. He reached up, involuntarily searching his own neck. Nausea welled up in his throat as his fingers encountered three of the slimy projections. His neck was festooned with them! No pain, no feeling of their presence, they were just there. Someone touched a cigarette to one of the brown projections. It reared up in slimy distress but did not fall off. The man grabbed it. Covered with mucus, it slipped from his grasp. Finally he took a rag and pinched and pulled with all his might, ripping it from the skin of the neck. Terrible pain. Blood flowed from the wound, running down the already wet neck, creating a fearsome crimson image. The blood was to flow sluggishly from the dime-sized gaping wound for another five hours. Most of the soldiers had scars, not from bullets but from these disgusting parasites. They were terrestrial leeches.

A month or so later, the shifting tides of war brought the men to another field of battle: the vast, flat rice paddies inland. Wading through the richly cultivated paddies brought no respite. Although the soldiers tied and taped their pants into their socks, they found aquatic leeches sucking on their legs and crawling in their private parts. Most Vietnam War veterans confess that they were at a loss to figure out how the animals penetrated their clothes.

This was a war of uncertainties. It was almost impossible to tell pajama-clad friend from foe; the citizens back home were rioting against the war. The only certainty about the war was the companionship of the leeches.

VAMPIRE WORMS

Leeches, like earthworms, are annelids and are characterized by ring-like annulations dividing their bodies into compartments. Other wormy phyla do not have these compartments. Earthworms and their close relatives, the salt water polychetes, have random numbers of these segments (up to a hundred), but leeches always have thirty-four (or thirty-three, depending on who is counting). Odd how every one of the six-hundred-plus species in this class has exactly the same number of segments.

There are two round suckers, one at each end. The three-lipped mouth is at the center of the anterior sucker. Four or more eyes, like black dots, lie on the dorsal surface near it. The eyes cannot define images but can differentiate light from dark—a pale human neck from the dark jungle.

The groovy skin is permeable to oxygen, but in order for this life-sustaining gas to slip through, the surface must be wet. The most obvious solution for leeches is to live in water. There are shark leeches in the ocean, turtle leeches in lakes, and fish leeches in both. But leeches of several families have evolved the ability to survive in the very humid air of rain forests. A few can climb trees and drop onto one's neck. These leeches can use an elliptical array of five pairs of eyes that can "see" in all three dimensions and sensory organs that can sense vibrations, carbon dioxide, and heat from an approaching warm-blooded animal. They have been observed to point the anterior sucker toward the source in anticipation of an attack.

The tropical terrestrial leech, *Haemadipsa zeylandica*, festooned the tropical trees and dropped onto the backs of U.S. soldiers in the rain forests of Vietnam.

The leech-human interface is complex. The leech cannot survive without a blood meal to provide the protein and energy to produce the dozens of eggs that might ensure the production of a new generation. But the wound produced by a leech can become secondarily infected, endangering the human host. Who wins this titanic battle between parasite and host?

The leech feints with its first blow, the lack of host specificity. Any "sensible" parasite will associate itself with a particular host, keeping its source of sustenance alive for as long as possible. But the leech chooses a cruder method. It attacks *any* potential source of blood. It chooses to forsake host specificity and sacrifices the ability to meld with its host on a physiological level. But this brutish interaction can cause the host to die. No problem. Like any temporary parasite, such as a mosquito, it tarries with its victim only long enough to sip its life-sustaining bloody meal. As long as the human population can recover from leech attacks, there will be meals aplenty for the next generation.

But it needs more than a moment to fill its distensible gut. There is evidence that, over eons of sucking blood, leeches have evolved saliva that contains more than the anticoagulant of the usual drop-on, drop-off blood suckers. The multifaceted saliva contains an anesthetic along with the anticoagulant, and a substance that dilates the blood vessels. So the leech sucks away painlessly and effectively. After its bloated body is engorged with blood, which can take more than twenty minutes, it drops off the host. For days it excretes the plasma of the blood, needing only the richness of those perfect capsules of food, the red blood cells.

By forsaking the protection and transportation provided by a more long-term relationship, the leech enters into an ephemeral interaction that is a bridge between parasitism and predation. Most land leeches have crossed the line and are predators on earthworms and other small animals that they swallow whole.

A DAY AT THE LAKE

You have finished your swim in the lake and are toweling off a brown "twig" on your calf. It won't come off. Horrifyingly, it undulates. *A leech!* You stare, fascinated. It begins to swell with your blood. Here is where fascination ends and revulsion starts. You need to fight back, to rip the leech off. The inevitable bloodbath follows. It is hard to grasp the leech. You squeeze frenziedly, but its slimy body is tough and slippery. Sated, it pops plumply out of your hand to lie in the underbrush, digesting your blood for weeks or months. Then, with the energy obtained from your blood, it produces dozens of eggs wrapped in the mucus that is its trademark. Its life cycle is complete.

Leeches are not picky. A turtle leech will latch onto a swimmer if it encounters one. Once it has crawled over its host and finds a good spot, the leech saws through the skin with three hundred tiny teeth (one hundred on each lip). So sharp are they, and so thin, that they stimulate few sensory receptor cells. Pain is a function of the number of nerve impulses reaching the brain at the same time. Sharp and thin means that few nerve cells will be touched—few sensory impulses, little pain. But having resolved the issue of detection, the worm must remain on the host long enough to fill the spacious, distensible gastric and intestinal pockets (cecae) extending from its gut. As if supersaliva is not miraculous enough, many leeches farm mutualistic bacteria, *Aeromonas hydrophila*, in the gut to help digest the complex blood. Miraculously, these intestinal bacteria grow in perfectly pure cultures, resisting competition by producing an anti-biotic that kills other species.

The bloated worm falls off the host to spend as many as nine months gestating its eggs. Leeches are hermaphrodites; each worm contains both sexes. Cross-fertilization with another leech is accomplished through the use of a penislike organ, or the sperms are simply deposited near the female gonopores. Eggs are released to be fertilized as they pass out of the body, or they are internally fertilized in the uterus. Several of the ringlike segments become swollen and differentiated from the others. This ring, the clitellum, produces mucus that envelops the eggs in a cocoon. The eggs hatch to produce miniatures of the adults.

BAD HUMOUR

The ancients believed that the body contained "humours," invisible substances that, if unbalanced, produce disease. For proof they pointed to the manifestations of the unbalanced humours in sick people: an overabundance of green phlegm or black or yellow "bile," or swelling induced by too much blood in an organ. The resolution of this problem was to reestablish the proper balance. To remove excess blood or to rebalance the humours, the sick person was "leeched." The medicinal leech, *Hirudo medicinalis*, was placed on the body. As if trained, it attached its anterior sucker and made an incision, injecting saliva into the wound. All watched as it became more and more turgid, swelling with the patient's blood. After a while the sated leech dropped off. The "cure" was accomplished. No matter that the condition became worse or the victim died, the theory was fixed—despite the data. Survival rates in the nineteenth century were as useless as body counts a hundred years later in Vietnam.

One tends to think of leeching as an ancient practice. A few years ago I was wandering through the Spice Market in Istanbul. A gallon jar full of dirty water stood conspicuously on a low table. Near the surface was a pale brown, fragmented, writhing band. A few of the flat, brownish "fragments" detached themselves, undulating through the water. With a tangible thrill it struck me that this was a jar of leeches

PLATE 19

A. TERRESTRIAL LEECH, *Haemadipsa zeylandica*. To 2 inches, light cocoa brown, three brown longitudinal lines; sharp annulations, scattered white pointed tubercles; anterior crescent of five pairs of eyes: respiratory swellings near rear for gas exchange; can detect vibrations, carbon dioxide, heat. On leaf; anterior sucker extended, sensing approaching victim.

B. INTERNAL ANATOMY OF LEECH. Anterior sucker leads to gray pumping pharynx and stomach with fingerlike gastric cecae, followed by more extensive intestinal cecae. All cecae become engorged, store blood for weeks, months. Anus opens above posterior sucker. Dotted circles, testes; long, banana-shaped organ on left, ovary; inverted L-shaped organs near front are male and female openings.

C. MEDICINAL LEECH, *Hirudo medicinalis*. To 3 inches, brown with darker streaks; thirty-four segments; mouth in oral sucker at thin end is sucking lymph from swelling around suture. Swollen part of finger above leech; most of leech swollen with blood, but visible segments at anterior signify leech is not fully fed; not ready to drop off.

D. CUPPING CUP. 2 inches high. When heated with match, some warm air escapes; cup is then rapidly pressed against human back, creating vacuum; when removed abruptly, leaves red welt.

E. Y-SHAPED WOUND made by three-lipped leech; teeth so sharp as to cause little pain; saliva contains anesthetic, anticoagulant. Area around wound swollen with lymph.

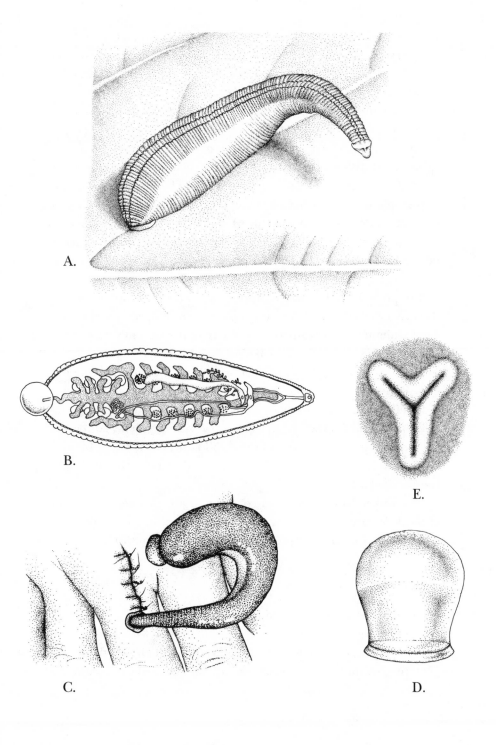

A.

B.

C.

D.

E.

sold for their curative properties—just another product for sale in the market. The stall was unattended. Furtively, I unscrewed the top and put my hand into the tepid water. None of the leeches attacked. I imagine that they must be literally placed on the skin to be effective.

The leeching theory was extended to encompass any means of sucking out the bad humours. Medical science in the nineteenth century produced a "high-tech" technique to remove the invisible humours, rendering the disgusting leech, with its attendant lesion, obsolete. Instead of "leeching," the patient was "cupped." The functional surgeon of the time was the barber. With the traditional shave and shampoo came another service, "a cure for what ails you." The barber placed thick, greenish glass globular cups on the naked back of the afflicted person so that a forest of globules protruded from the skin, similar to the masses of needles applied during acupuncture.

The procedure: The barber heated the inside of a cup with a candle, driving out the air. The warm cup was then clapped onto the skin. The resultant vacuum caused it to adhere to the patient's back. After the proper amount of time, the barber pulled off each cup with a loud pop. Presto! The bad humours were removed. Evidence proving the effectiveness of the cupping process was provided by the angry red welts left on the skin, testimony to the power of the cups. Lest this be relegated to olden times, I dimly remember sitting with my father in a barbershop in Brooklyn when I was a child during the Great Depression, witnessing this process. Evidently it is still practiced in Russia and the Middle East.

In a souk in Israel I found one of the cups jumbled among ancient artifacts. Instantly recognizing it for what it was, I bought it for a few shekels. An illustration of the cup can be seen on plate 19.

A MODERN MIRACLE

A surgeon reattaches a severed finger to a patient's hand. Using precise microsurgery, the surgeon rejoins blood vessel to blood vessel, nerve to nerve. This extraordinarily delicate surgery finished, the surgeon reaches into a box and removes *a leech.* The medicinal leech has found a new role.* The wound inevitably swells with blood and lymph flowing from damaged tissues, and the leech sucks them out more delicately than any modern device. The surgeon places a bib of gauze around the swollen spot, and the oral sucker is placed over a hole in the bib. If the leech does not cooperate immediately, the surgeon makes a tiny bloody hole with a pin to inspire the animal.[20]

Calling all entrepreneurs: my million-dollar idea of the day. Open up a leech farm. But competition is not as sparse as you may think. A "big" farm (a few rooms) in England competes with an esoteric Lithuanian leech farm. When I was researching this chapter, I found to my amazement that a leech farm is located only a few miles from my home. I must drop by someday to ask them the selling price of each medicinal leech. What with the high price of medical procedures these days, I bet you can sell each leech for a fortune.

Many thanks to Jack Ferrera, Spec 5, veteran of the Vietnam War, for serving as consultant for this chapter.

---

*Two substances in leech saliva have useful medical applications. An anticoagulant, hirudin has been used for years. The local anesthetic that allows the leech to bite painlessly has been synthesized. The hirudin is so effective that the mean time of postbite bleeding is 490 minutes under controlled conditions.

20

ODE TO A COCKROACH

La cucaracha la cucaracha
Te presento la cucaracha
La cucaracha la cucaracha
Ya no puede caminar
Porque le falta
Porque no tiene
Una pata para andar.

THE WILD LIFE IN NEW ORLEANS

At this writing, unique and marvelous New Orleans, with its "easy" streets, exotic, sinful French Quarter, and redolence of jazz and jasmine, is smothered in filth, a monument to nature's wrath and mankind's incompetence. But it will arise like a phoenix. Its streets will inevitably teem again with its earliest returning inhabitants—cockroaches.

WILDLIFE IN NEW ORLEANS

Walking home from the lab late at night was an adventure. Huge living missiles whizzed past my head; crunches emanated from underfoot. Hot, humid New Orleans, an early warning of what might happen as a result of global warming, was a breeding ground for all sorts of night-dwelling animals. It didn't freeze in winter. Its warmth-loving, huge live oak trees, laden with exquisite wreaths of Spanish moss, shed copious leaves, providing moist, dark havens for underground things that crawled out of leaf-strewn muddy pools of rainwater and hid during the day. Temperatures hovered near humidity levels of 90 percent.

We purchased a gigantic old air conditioner. It creaked and rattled day and night. The relatively cool, dry air and a 24-inch TV made our apartment a haven for the 1960s hippie couple living the impoverished life of students downstairs.

Mary didn't study except for the week before finals, when she made a benzedrine-fired, twenty-four-hour-a-day total immersion effort to learn the semester's coursework. *This was that week.* Taking a break from her studies, she arrived at our doorstep, visions of Sid Caesar's shenanigans in her head. She knocked; I opened the screen door, noticing that she was wearing a fashionable peasant blouse. Suddenly a huge flying cockroach landed on her bare shoulder. She freaked! The slender thread of sanity broke. I dragged her upstairs and flung her on the couch. Wise wife sat on her abdomen trying to soothe the shrieking and writhing. I ran around wringing my hands until it occurred to me to get her husband. Together all of us calmed her. The result: she scored A's on every test. The annual ambulance took her away at the end of the week, and I became fascinated with cockroaches.

PERFECTION IS A MATTER OF TASTE

The cockroach the cockroach
Allow me to introduce the cockroach
The cockroach the cockroach
Can't walk anymore
Because he's lacking
Because he doesn't have
A foot with which to walk.*

I was wary of starting a new project. Investigations of parasite taxonomy require investing total energy for as long as it takes. With this

---

*The folk explanation for these lyrics: The Mexican revolutionary hero, Pancho Villa, had a carriage that was always breaking down on his endless journeys around the desert. His troops named the bulky carriage "la Cucaracha."

in mind, I carefully searched for the appropriate host animal. It would have to be abundant and easily obtained, and as full of parasites as possible. The cockroach beckoned. The dark streets abounded with them. But to my surprise, they were not so easily obtained. Waving long, sensitive antennae, with eyes that efficiently sense movement, they are elusive when approached from the front. At the posterior, longhornlike cercae sensitively test the air for any movement. Each of my lunges initiated instantaneous flight into a dark haven. I took to running madly through the streets waving a net futilely. It was embarrassing.

Finally wise wife found the solution. When I arrived home for dinner one evening, I was greeted by two cute baby girls—and two or three upside-down paper cups weaving rapidly along the floor. Unsuspecting roaches, fat and listless from sating themselves on crumbs fallen from kiddie meals, were easy prey. Wise wife sneaked up on them and plopped a cup on each one. Stunned, at first they felt secure in their dark paper prisons. After a moment they moved—and carried the confining cups with them. I slid an index card under each cup, turned it over, and plopped its contents into a jar for tomorrow's investigation.

At the lab I anesthetized them with carbon dioxide until they were dead. Then, just in case, I cut off the head, legs, and wings. Beheaded female roaches often run away and end up in a crevice, inaccessible. Then in one convulsive reflex, they disgorge an ootheca, an elongate, oval capsule with about twenty partitions, each containing an egg. One must be very careful lest their last revenge produces, instead of a reduction in their population by one, an additional twenty.

Then the adventure began. I removed the upper carapace and exposed the grayish gut. Tugging gently, I freed the gut in its entirety. In a dish of saline solution, it was divisible into three segments: a foregut of swollen crop and hard gizzard, a midgut of stomach and intestines, and a hindgut of large intestine (bowel) and rectum. The midgut was delineated by finger-shaped gastric cecae. Massed white threads of the excretory system demarked the beginning of the hindgut.

The fun of investigating the gut under a microscope is not knowing what symbiotic inhabitants lurk in its dark, moist interior. I teased the foregut apart with pointy delicate forceps. Nothing. Then the midgut. *Bingo!* Thrashing through the gut contents were three kinds of sausage-shaped roundworms with long, tapering tails, revealing that they were in the same superfamily as pinworms. They were huge—visible even to the unassisted eye as writhing dots.

Fascinated, I stared through the microscope at the twitching, transparent worms so intently that my eyes began to tear. Needing respite, I glanced at the walls of the midgut, expecting the bland, wrinkled normal appearance. To my surprise the walls were festooned with "bowling pins." Unmoving, just projecting into the lumen of the intestine, they might have been a metastasized cancer. But it suddenly struck me that these were gregarines. Their relatives are dangerous parasites.

Overwhelmed with my discoveries, I considered what lay before me—the hindgut with its juicy colon full of that parasite favorite, feces. With great anticipation, I slit open the sausagelike tube. Revealed was a veritable jungle full of wild animals. Huge *Paramecium*-like ciliated protozoans swam majestically through the circle of light that was the microscope field. Amebas crawled along the edges of the gut. Visibility was impaired by a cloud of bumbling, erratically swimming flagellated protozoa of many species. Mixed in with the contents of the hindgut were shelled ovals that I recognized as the eggs of the roundworms, and tinier capsules that were cysts of the amebas—eggs and cysts to be released with the feces. Other roaches, not inhibited by human values like disgust, would eat the moist feces of their brethren and become infected in turn.

Altogether satisfying, the abundance of life in the roach gut suggested further investigation. I knew that the taxonomy of the symbionts inhabiting the gut of roaches had been studied in the early twentieth century. But another view using modern techniques might clear up some confusing discrepancies.

RATS, ROACHES, AND RESEARCH

Fond as I am of rats, I had found a more magnificent subject for investigation, one of the three common species of cockroaches inhabiting dark spaces behind things, masses ready to boil out when uncovered. Smallest is the German roach, *Blatella germanica*. Larger is the blackish "water bug" often found in basements, the oriental roach, *Blatta orientalis*. Most magnificent is the American roach, *Periplaneta americana*. (But sometimes, behind a wall or under a rug in Florida, it is possible to run across the ultimate roach, the mind-boggling, three-inch-long Palmetto bug, *Eurycotis floridana*.)

Each kind of roach is inhabited by species-specific symbionts. Down south and in warm buildings and sewers everywhere, the American roach, with its own internal cargo, holds sway. When not buzzing through the air in New Orleans, it crawls rapidly on sidewalk or floor, bringing near-hysterical responses from housewives and husbands.

WHAT'S EATING ROACHES

The pinwormlike nematodes in the cockroach gut flaunt sexy elongate tails, but sexual selection is elicited by stimuli other than visual in the dark confines of the roach gut—most likely by those sexy, perfumelike pheromones, for roundworms lack eyes. My favorite species, *Hammerschmidtiella deisingi*, vies for the longest name of any invertebrate.

The large *Paramecium*-like ciliates that I saw were *Nyctotherus ovalis*. Like swimming salamis, they moved in straight lines, living torpedoes gliding with unwavering trajectories through the contents of the hindgut.

The giant bowling pin-like gregarines in the midgut are remotely related to malaria parasites. They don't move, but they are more dangerous than the other species. They are parasites. While many of the other (commensal) symbionts in the gut are content to subsist harmlessly on the ever-available nutrients in the intestinal fluid and feces, these animals, the unmoving gregarines, penetrate the gut wall when they enter the body. Soon after asexual reproduction in the cell, they burst out and attach to the intestinal wall, becoming the bowling pins.

Somehow these identical bowling pins are able to figure out which is male and female. They line up, the male attached to the posterior of the female, in a process (the ultimate boon to scrabble players) called syzygy. Then, wrapped around each other, they encyst and divide into many infective cells (spores) that pass out in the feces to be eaten by another roach. Close relatives (coccidia) cause disastrous epidemics in farmed animals like rabbits and chickens.

It has been shown that three species of these innocuous-looking bowling pins are distributed within the narrow confines of the gut according to the pH (acidity) of consecutive sections of the upper intestine. Their ecology makes them useful models for studying their human-infecting relatives. But that was not my concern. I was interested in how to distinguish each species, since all have just about as many surface characteristics as the bowling pins all resemble. (At that time DNA-based methods of classification were not commonly used.)

The clouds of flagellum-bearing protozoa in the hindgut bore some species unique to cockroaches and one or two families containing parasites similar to human parasites. One of these families contains the aptly named human vaginal parasite *Trichomonas vaginalis*. It causes a venereal disease dangerous to women but is relatively harmless to men. Untreated infections can destroy the lining of the fallopian tubes, preventing the passage of the egg to the uterus, causing sterility. It is hard to detect and often discovered when it is too late. Culturing populations of *T. vaginalis* is difficult, so the species from cockroaches can be used to study the physiology of these flagellates. It has been estimated that 25 percent of the world's women harbor this parasite. (Not to worry. In this era of newly discovered venereal diseases, physical methods of contraception that protect you from other STDs protect you from *T. vaginalis*.)

THERE IS NOTHING AS DISGUSTING AS A ROACH, EXCEPT A COLONY OF ROACHES

Generations of students in my classes have benefited from the ever-generous roaches. I have cultured them for over forty years in aquarium-like containers in the classroom. Hundreds of reluctant hands have reached in to obtain a roach to dissect.

Let's get this straight. There is nothing as disgusting as a roach— except a colony of roaches. In addition to the "boiling-out" phenomenon when you pick up a paper towel under which a dozen have been hiding, there is a characteristic stench emanating from the colony. To assure myself that they do not escape from the completely covered container, I smear the upper inch of the glass wall with petroleum

PLATE 20

A. AMERICAN COCKROACH, *Periplaneta americana.* To 1¼ inches, brown with yellow markings on thorax. Legs spiny; wings folded on dorsal surface. Two "horns" at rear, cercae, can perceive vibrations.

Organisms below are some of many species found in cockroach gut. Sometimes, when gut is dissected, hundreds of protozoans are visible in fecal fluid.

B. COCKROACH ROUNDWORM (NEMATODE), *Hammerschmidtiella deisingi.* To 1/16 inch, visible to naked eye; female has long, pointed tail; note eggs in uterus. Pharynx, gut in front; anus opens at tail. In roach midgut. Related to human pinworm.

C. COCKROACH GREGARINE, *Gregarina blattarum.* To 1/16 inch, visible to naked eye; featureless, except for nucleus. Parasitic protozoan; kills roaches because asexual stage bursts out of gut cells, destroying them. Lines up for reproduction: female hangs from gut wall, male (active cell) attaches to her posterior (syzygy). In roach midgut.

D. HINDGUT (RECTAL) COMMENSAL PROTOZOANS
  (1) COCKROACH AMEBA, *Endamoeba blattae.* Microscopic; ring of dots around edge of nucleus; amorphous. Encysts in feces. Genus confined to roaches, related insects. Related to human dysentery ameba, *Entamoeba histolytica.*
  (2) COCKROACH FLAGELLATE, *Retortamonas blattae.* Microscopic; two flagellae; nucleus at left contains karyosome (dot); encysts in feces. Usually oval; one commensal species uncommon in humans.
  (3) COCKROACH FLAGELLATE, *Tritrochomonas blattae.* Microscopic; two flagellae. Closely related to human reproductive tract parasite, *Trichomonas vaginalis* (shown); 4 anterior flagellae; one trailing flagellum runs along undulating membrane; cause of very common human STD, and *Tritrichomonas foetus,* cause of abortions in cattle.

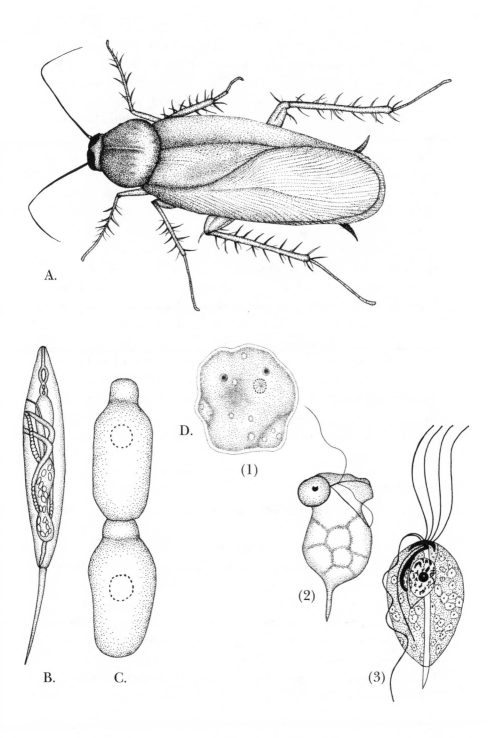

A.

B.   C.

D.

(1)

(2)

(3)

jelly, creating an impenetrable barrier. The roaches climb rapidly up the glass, but when they reach the petroleum jelly, they slide backward, leaving tiny trails running down the wall. Nevertheless, the biology building is infested with roaches—and I am blamed. I show the petroleum jelly barrier to my accuser and proclaim my innocence, to no avail.

One year my colony died. I resorted to collecting roaches from the building. A friend brought me a few from the ladies room, but I needed many. My colleague, grumpy, scholarly old goat, told me that Rutgers University had a whole building devoted to raising roaches. Desperate, I drove there with my glass containers and entered the dark, silent confines. Out of the gloom appeared a man who reminded me of Igor (Eye-gore) in a Frankenstein movie. He kindly gave me several handfuls of roaches, including the three-inch-long Madagascar hissing roach, *Gromphadhorina portentosa*. Back at the classroom, a student reaches into the aquarium-like container and grasps one, scaring it. The huge animal compresses its body, rapidly releases internal air from modified breathing holes called spiracles, and emits a blood-curdling hiss. It is hard enough to screw up your courage to grab the huge insect, but the loud hissing sound is such a shock that many a student drops the roach and runs, leaving a three-inch-long wild cockroach to wander around the building.

SPANISH DANCER

Each of the classroom colonies of American roaches contained about fifty insects.

The idea was to obtain a roach, terminally anesthetize it with carbon dioxide, and dissect it, revealing its intestinal symbionts. To examine them, the student glued a gob of intestinal juices onto a slide and painstakingly stained it for later examination. Required: two slides with identified symbionts attached to an elaborate paper to be handed in at the end of the semester. The purpose of this paper was NOT to train the students to be experts on the cockroach intestines. We capitalized on the miraculous contents of the roach gut to teach the students what they didn't know—*how to function*.[22]

The process of preparing the slides is more difficult than you may think. The first effort is to remove a roach from the squalid colony. This challenge teaches many things. The students are repulsed. The task seems impossible. Some unemancipated women decide to employ a time-honored method—they smile at the young man next to them and request assistance. No matter how independent a woman, there are times when she is not adverse to using charm. In this moment of desperation, a sensible woman asks the "guy next door" to get her a cockroach. He is equally repulsed (as I am), but his male honor is at stake. Nothing is as powerful as the need for a youth to respond to a challenge to his manhood. The young man reaches into the petroleum jelly-smeared tank and scrabbles around in the sawdust at the bottom, a look of disgust distorting his face into a weird grimace. He succeeds in placing the roach into the jar she is holding. She claps on the cover and runs to the tank of carbon dioxide, placing the hose inside. Occasionally, in her nervousness, she turns the gas on too hard and the roach is blown by the blast across the room. People duck, shrieking and bellowing, and the roach is lost. The gallantry of the young man is then further tested. She needs another roach.

Increasingly women do not resort to feminine wiles. Aware that they are no more turned off by roaches than are men, they reach into the tank. One such occasion is burned into my mind. A delicately pretty young woman, face stony with determination, reached into the tank, causing the usual boiling out of roaches from under their dark havens. She grasped at one. It ran up her arm toward her vulnerable armpit. A gasp! Determination gone, she uttered a nasty word and flung the roach onto the floor. Enraged, she stomped on it . . . once . . . twice . . . many times. All she needed was castanets. The sound of her heels was as staccato as any Andalusian dancer.

21

BATS, BUGS, AND BLOODY BITES

THE EXPEDITION WAS A GREAT SUCCESS. My colleagues returned home with small vials of specimens suspended in the subzero cold of liquid-nitrogen-filled stainless steel canisters. Brought back over thousands of miles from uncharted jungles of South America, they were bat salivary glands.

Why bat salivary glands? They must be important. The work of the distinguished leader of the safari was subsidized by a number of federal agencies. Or does research on bat salivary glands qualify this project for the Golden Fleece Award presented by former senator William Proxmire for scientific projects he deemed "a waste of taxpayer money"? (Thus did a powerful politician seek to bypass the normal scientific oversight system controlling funding of research.)

Bat glands are uniquely suited to determining taxonomic distinctions. The glands of fruit eaters are distinct from those of carnivores, and both are different from the glands of insectivores. Under the microscope aggregations of salivary secretory cells form flowery fields as distinct as those in a garden. These cell clusters give insights as to the radiation of species from common ancestors, making them an important source of information about their evolution.* Bats are often the only mammals indigenous to some islands. Other mammals we find were either purposely introduced (like mongooses introduced into Jamaica to kill snakes) or disastrous accidentals, like pigs, goats, cats, and rats. Bats have not been eradicated like dodos because they are not as vulnerable, so it is possible to trace lineages from island to island in a kind of Darwinian description written in spit.

One of the investigators in the group was an experienced trekker, but she had not camped in the jungle before. One day she walked into the bush to relieve herself. She knew enough not to squat in the unfamiliar undergrowth and used a plastic device that allows women to urinate standing up.[†] But she was in such a hurry that she didn't change her shoes. In the comfort of the base camp, she had replaced her stuffy boots with cool flip-flops. Bereft of shoes and socks, she was vulnerable.

A few days after her return home, she noticed a small lesion under a toenail. It expanded into a pus-white, painful mass. The doctor had no idea what it was, so he did what ignorance dictates. He removed the nail and scraped out the white substance that had been exerting painful pressure. All of this created considerable discomfort, which she vented, it is rumored, with great vehemence.

The lesion was caused by a *FLEA*. Not any flea, but that miniature monster, the chigoe, *Tunga penetrans*.[‡] And penetrate it does. Once she digs her way under the toenail, the female flea deposits up to 150 eggs. Eventually this hump-backed female Frankenstein swells up to look like a pea with a tiny head. Buried deep under the skin, she becomes distended with blood and eggs. Over time, her legs and external features become absorbed and she becomes a pea-size egg-laying machine. The larvae are released from a permanent pore in the host's skin to develop on the ground— there to wait for another victim.

---

*Nowadays these taxonomic and evolutionary problems are resolved by examining mitochondrial DNA and using other molecular techniques.

†Available in any camping supply catalogue. I don't know under what heading. According to the *New York Times* (Aug. 17, 2009, p. A6) it is pink and is called a "feminine urinary director."

‡This flea is also called the "jigger" because its painful presence causes the victim to "jig" from foot to foot.

Eventually the open hole under the nail becomes infected. Tetanus and gangrene often result. The toe rots off.

ISLANDS IN THE JUNGLE

Why mount an expedition to uncharted jungles of Suriname to get bats? The destination of this band of serious scientists was a series of flat mountaintops that were once contiguous. But unique tectonic action separated the mountains from one another, thereby separating their indigenous flora and fauna. They were far enough away from one another for evolution to occur in each separate mountaintop "island" of jungle. This Darwinian laboratory is reminiscent of the Galapagos Islands, but these islands are separated not by ocean but by jungle-covered lowlands. Unique species evolved, each particular to its own mountain-top community.

BAT SALVATION

My laboratory in Jamaica was a perfect headquarters for another salivary sojourn. Our leader was interested in the glands of the Jamaican fruit bat, *Artebeus jamaicencis*. All he needed was one or two. I was able to provide them because I knew of a "bat cave." Not any batty cave, but one whose population I had saved.

Every year I bring my marine biology class to Christopher Cove on the north coast of Jamaica, twenty minutes by boat from the lab. When approached by sea, the first distinguishing feature is a black hole in the limestone cliffs bordering the cove. This mysterious dark scar in the yellowish wall is Bat Cave, named for the colony of fruit bats suspended from its ceilings. But this is not a replica of Carlsbad Caverns where thousands of bats hang from the roof of the cave, barely visible in the dim darkness as an undulating, indistinct, upside-down carpet high above. This is a pipsqueak, scarcely thirty feet across and twenty deep.

The students snorkeled to the cave and entered underwater, a sometimes formidable task when the sea is high. But the scariest part was when they stood up inside, with the bat-shrouded ceiling only three feet above their heads. Disturbed, the bats would flutter about among the students, emitting their characteristic, barely audible squeaks. The squeaks produced by the women in the class were not barely audible. They were loud enough to flatten the ocean. This experience, while tangential to the purposes of the marine biology course, is remembered to this day by the fifty-year-olds who took the course decades ago, even if the fishes and corals are forgotten.

One year a man appeared on the cliff above the cave. He held those characteristic candles whose purpose is to eradicate bats. The smoke is laced with cyanide. To my horror, he was making his way down the face of the cliff with his candles. Treading water, I removed my snorkel to talk to him. Water entered my mouth. In a gargling voice I harangued him with typical Jamaican banter, which requires telling a few jokes before getting to the subject. He was the owner of a dude ranch bordering the sea, and someone had complained about the bats. I discussed their value to the tourist industry and their role in the fruit industry as carriers of pollen, all the while treading water, trying to appear jolly, the appropriate flippant remarks gurgling from my drowning mouth. At last he left, carrying his candles with him. To this day the cave contains its precious population.

ITCHY ARACHNIDS

Not satisfied with a couple of fruit bats, the aforementioned intrepid researcher wanted the glands of the comparatively rare fish-eating bat, *Noctilio leporinus*. This entailed capturing the bats as they fished in a cattle-watering pond on a local farm. At midnight we set up our mist nets on high poles. So spidery was their mesh that their filmy folds were virtually invisible to human and bat. We sat down, uncomfortable in the stubble of the newly mown field, making ourselves as inconspicuous as possible to avoid disturbing the bats.

Finally, silhouetted in the dim moonlight like a Halloween poster, the first bats appeared. They skimmed over the pond, touching lightly down on the still surface, leaving scarcely a ripple. Suspended from their claws were their prey, tiny minnows. Eventually some of the bats flew into the filmy nets. It was then our job to remove them. In a flashlight beam, a struggling bat is intimidating. Its blunt face is hideous, and the threatening grimace of pointed white teeth in its snarling mouth makes the animal appear formidable. No matter that it was only three inches long, it was scary. I donned pigskin gloves and grabbed the bat in a smothering grip. It struggled fiercely, bound by the enshrouding mesh. I was supposed to carefully unwrap the bat so as to conserve the net. Damned if I would be neat. I ripped the animal from its cocoon of threads and, though it was festooned with fibers, I didn't care and plopped it into the container. Stunned, it sat there in the glare of the light until I popped the cover on. The big hole left in the net when I tore the bat out didn't impair its usefulness, and we caught ten bats that night.

A few days later I was awakened by fierce itching around my waist and along the bottom edges of my underwear. Mosquito-bite-like welts traced an erratic row along my belt line. What mosquito would wander up my pants to bite along the constrictions of my belt and underwear? All at once it hit me. Chiggers! They are well known for their habit of climbing up the huge columnar mountain that is the human leg, only to be stopped like a dam stops salmon migrating upstream, by the tight band of underwear. There they feed, the bites apparent a day later as a row of itchy welts. Sitting in stubble was our undoing; it provides ideal habitat for these unusually malicious ectoparasites.

Chiggers, species of *Trombicula*, are not insects like fleas. They are arachnids, along with spiders and ticks, and are classified as mites. So tiny are the members of this order that some species are called dust mites, and when inhaled they can cause allergic reactions. Along with the others of this class, mites have eight legs (in contrast to fleas, lice, and all other insects, which have only six) and mouthparts that consist of paired, stiletto-like structures so sharp that slicing the skin does not cause pain. Another pair of mouthparts holds the wound open while sucking up predigested skin cells.

Chigger feeding differs from the classic pattern. They do not suck blood. It is your skin cells they are after. Their horrendous feeding technique makes mosquitoes seem benign. Chiggers secrete enzymes that digest your skin. Then they suck up the potpourri of dissolved cells mixed with lymph. But their nastiness is not finished. Somehow they have evolved the capability to use your own body to help perform their ghoulish work. Their saliva hardens your cells into a drinking straw to help them to suck up their cocktail of dissolved cells and intercellular fluid. After their awful meal you are left with these discarded microscopic cellular tubes as souvenirs.

The adult chigger mite is huge for its kind. Its red body can reach a sixteenth of an inch. Males wander around, and when the urge strikes, they deposit a bag of sperms, a spermatophore, on the ground. A female comes along and sits on it, inserting the spermatophore into her genital pore. After fertilization, eggs are laid. They hatch in a few days and the larvae wander around looking for a host. They are very active, as if they know they will die after two weeks if not able to find a host. It is the larvae that attack humans. Though they are not fully developed, their six walking appendages are adequate for climbing up the leg of an unsuspecting animal. Chiggers are not particular—any mammalian host will do.

Some humans do not exhibit symptoms; their immune systems do not respond to the proteins in the mite's saliva. Sometimes the response is delayed. Once, in Belize, a class arrived at the lab. Most, but not all, had an uncomfortable, distracted look in their eyes. The reason soon became clear. Previously they had performed community service at a little mountain village by clearing stubble from a soccer field. On close inspection of coeds wearing bikinis the next day, I saw the characteristic lines of bites. To their credit, those bearers of welts were so involved with their field trips to coral reefs that they didn't seem to mind the insanely itching bites. My hat is off to them.

TSUTSUGAMUSHI

To add to the misery they convey, larval chiggers are vectors of the mellifluous deadly disease tsutsugamushi, or scrub typhus. Originally described in China two thousand years ago, this rickettsial (bacterial) disease killed many soldiers in the Pacific theater in World War II. A lesion appears at the site of the chigger bite. It slowly spreads until it is quarter size, then it festers in the middle and the cells disintegrate. As if that is not enough, other symptoms are delirium, enlarged spleen, and, rarely, deafness. Before antibiotics death occurred in up to 60 percent of those bitten.

BACK TO THE BATS

Flying slowly over a field in the pitch blackness of the night, the vampire detected warmth emanating from the hot blood and warm body of a large lump almost invisible in the grass. Unheard by us, high-pitched sonar waves emitted by the bat returned from the object, verifying that this was a pig. As we watched, it landed on the ground a few feet away from the pig. The bat crawled amazingly rapidly on the ground, taking seconds to cover the distance that separated it from its sleeping porcine prey. It rapidly mounted the animal, crawling like a black, angular blob across its body. Still the pig did not wake up. "Sniffing" with infra-red detectors in its nose, the bat located a plexus of capillaries, their combined warmth signaling that this was a blood-rich area. Using its famous canine teeth (which are not hollow and lack enamel so that the vampire can hone them to exquisite sharpness), it made a quarter-inch, virtually painless incision. Anticoagulant-rich saliva poured into the wound. Blood sluggishly seeped out. The animal crouched over the wound and *lapped up the blood.* The ghastly process was taking too long. To hasten it the bat inserted its tiny tongue into the lesion and slathered it around, forcing the blood to flow more rapidly. The tongue has long lateral grooves, carrying the blood into the mouth faster and faster. Few chewing teeth are located in the depths of the mouth—few are needed for the bat's liquid diet.

Sated with blood, the bloated bat urinated automatically, releasing a flow of dilute urine. The watery component of the blood was released in the urine, lightening the vampire's burden and leaving a thick soup of protein-rich cells in its stomach. The bat staggered off the pig and crawled drunkenly along the ground. Suddenly it crouched. Using its huge muscular thumbs, it leapt a yard into the air and seamlessly switched to flying mode, disappearing into the dark night, a whisper in the blackness.

A dim opening appears, a deeper black in the black night. It pulsates almost audibly with the breaths of hundreds of the bat's brethren.

Fluttering into the fetid darkness, the bat uncannily locates its family in the cave's twitching, undulating, furry roof. Each bat clings closely to its neighbor, all huddling together for warmth. Once ensconced, the bat defecates, its droppings melding with hundreds of years of deposited feces, forming inch-thick layers of guano on the cave floor.

A juvenile, unsuccessful after the night's hunt and barely surviving after two days of starvation, flutters over weakly. Desperately grasping at the carpet of contiguous fur lining the cave ceiling, it produces a flurry of writhing, disturbed movement as sleeping bats get out of the way. A toehold on the rocky, ridged roof is exposed and grasped. The hungry juvenile hangs upside down next to the male, secured by sharp fingernails. Then it nuzzles up toward the mouth of its engorged father like a puppy and, in a swift kissing movement, latches onto it. With a convulsive heave, dad blurches a bolus of blood into the sucking mouth. Blood means survival. Juveniles are often unsuccessful at finding the two tablespoons of blood that they require every two days.

A BAD REP

The vampire bat, *Desmodus rotundus*, is described in song and story as the ultimate villain. In most legends the bat morphs into an undead human who sleeps in a casket and arises with the moon to suck the blood of its human hosts with long fangs.

Here is the truth:

- Vampire bats do not have hollow fangs. Their canine teeth are long and sharp. They make an incision that makes it possible for the elongate slithery tongue to lap up the blood.
- Vampire bats do not prefer the jugular veins of the neck; any vulnerable hairy surface will do. Their incisors are scissorslike to clip the host's hair like the neat barbers they are. All this requires the host to remain asleep. If the host awakens and tries to remove the horrible, grotesquely grinning animal on its back, the bat will scamper around its host's body in a ghastly duel with death.

PLATE 21

A. VAMPIRE BAT, *Desmodus rotundus.* To 3 inches; front teeth (incisors) sharp, used to clip fur of host; long canines make 1/2 inch incision in skin. Laps up blood with long, narrow, grooved tongue (no sucking); lower chin has V-shaped callous that facilitates lapping; nose can sense warmth. Bat can differentiate between breathing patterns of horses and cows using large ears that permit object avoidance while in flight by detecting sonarlike bursts of squeaks. Three substances in saliva: anticoagulant, chemical that prevents red blood cells from sticking to each other, anesthetic.

B. JIGOE, JIGGER FLEA, *Tunga penetrans.* To 1/32 inch, smallest flea; burrows under toenail, lays eggs in white blister with central dot. Can jump only 6 inches, thus usually found on foot; releases one hundred eggs through dot that hatch on ground, then female dies, is absorbed or expelled. Eggs hatch into wormlike larvae that eat detritus.

C. FLEA ABDOMEN SWOLLEN WITH EGGS. Pea-size and shape; posterior bulge contains spiracles (nostrils), reproductive opening.

D. BLISTERS ON TOE. To 3/8 inch, white; dot contains flea's hind legs; spiracles; reproductive opening.

E. CHIGGER MITE, *Trombicula alfreddugesi.* Adult to 1/20 inch, has eight legs; six-legged larva is parasitic; velvety; red or yellow. Do not suck blood; dissolve skin cells, predigest them, suck up cell-enzyme mixture. Adults harmless; females lay fifteen eggs/day that hatch into larvae that feed on skin up to three days, drop off host, become eight-legged nymph, then adult in soil. Small front "legs" are mouthparts.

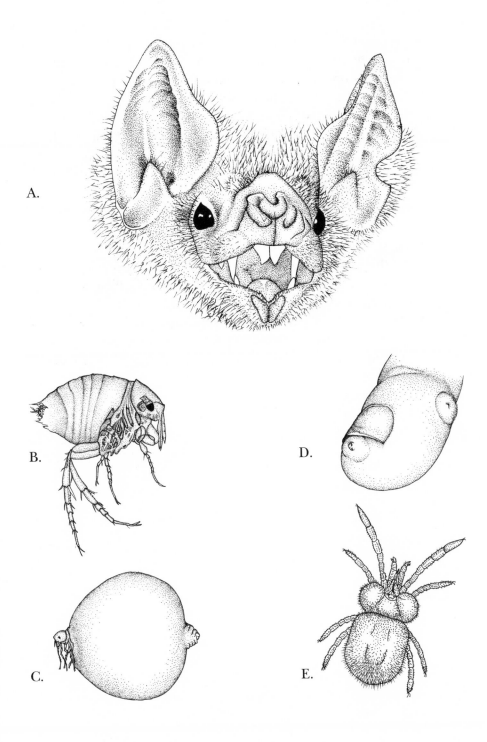

A.

B.

C.

D.

E.

- Vampire bats live in South and Central America. They have a high body temperature, and this limits their range to tropical and semitropical regions. With global warming, their range is increasing into the southern United States.
- Vampire bats rarely attack humans, and then only when they sleep outside. They don't seem to like the taste of our blood. They cannot enter bedrooms that have screened-in windows.
- Vampire bats do not remove a significant amount of blood. The victim is no paler after his/her bloody experience than before. How much blood can fit into the stomach of a three-inch-long-animal? About an ounce.
- Vampire bats can be vectors of diseases like rabies, but only if they have fed on a rabid animal. No cases of vampire bat-borne rabies in humans have been recently reported, but thousands of cattle die every year of rabies from vampire bat bites.
- The conversion of vast tracts of tropical forests into megaranches for Burger King and McDonald's has caused an increase of vampire bat populations (possibly the only tropical animals that are increasing in numbers).
- Teenagers, as depicted in the movies as vampires, suck only their family's fiscal resources away. They are not really monsters, they only seem to be.

22

LITTLE FLEAS HAVE LITTLER FLEAS

Little fleas have littler fleas
Upon their backs to bite 'em
And littler fleas have littler fleas
And so on ad infinitum*

WE WERE COMING HOME FROM A CAMPING TRIP. Our dog, Bandy—named for the shape of her legs—was on the back seat, sleeping on wise wife's extra skirt. Bandy was an apartment dog. Her limited life was devoted to destroying drapes, eating furniture, and defecating on the floor. Here was her opportunity to get into the great outdoors. Ecstasy. She ran around in the woods for the whole week of vacation. Evidently this paragon of apartment life had met another dog. Her pristine cleanliness was sullied.

We reached a restaurant. Wise wife put her skirt on over her shorts. After a minute or two, before we could order, she became upset and ran out, jumped into the car so that no one could see, and started flailing about. On her legs was a blurry mass of moving dots. I beat at the dots with a rolled up newspaper, hardly able to swat my minute

---

*The title of this chapter refers to the fact that it is possible for even parasites to have other parasites inside them. The oriental rat flea has a bacterial parasite inside it. This phenomenon is called hyperparasitism.

targets on her gyrating legs. Finally, she ripped off her skirt and ran madly around in the parking lot. The flea-flecked skirt was thrown into a waste basket and we made our itchy way home.

Over the years other incidents like this occurred. It became apparent that wise wife is attractive to fleas—even more attractive than the dogs whence they came.

Some women are just attractive—to fleas as well as men, and just maybe for the same reason. Estradiol, the female hormone, permeates a woman's skin, and she sheds potent molecules into the air. Can human males detect these minute molecules wafting by? Fleas can. Is that the secret of her universal attraction?

Everyone releases carbon dioxide from skin pores; some people release more than others. It has been shown that this product of human respiration attracts fleas, mosquitoes, and other biting insects. But we have no evidence that she produces more carbon dioxide than most. Clearly several factors combine to make her so attractive. Another suspect is the smell of perfume. Buried deep inside some brands of perfume lurk molecules of musk, a sexual seducer of mammals. Is it possible that a woman who wears this musky perfume is more seductive to fleas? I experimented. A surprise present, a new kind of (musk-laden) perfume, produced no change in the unwholesome interest of the fleas. (Never has one of my scientific experiments yielded so much enthusiasm from wise wife.)

Wise wife may have to resign herself to being so attractive.

The combined effects of Nero and Kubla Khan, of
Napoleon and Hitler, all the Popes, all the Pharoahs,
and all the incumbents of the Ottoman throne
are as a puff of smoke against the typhoon blast
of fleas' ravages through the ages.

—B. LEHANE

NURSERY RHYMES

Ring around the rosie
Pocket full of posies

Ashes, ashes
All fall down

Nursery rhymes innocently repeated over the generations to two-year-olds are often hideous half-memories of ancient evil disasters, death, and calumny. Under the English monarchy in the seventeenth century, dissent was punishable by death. Nursery rhymes had hidden meanings. Georgie Porgie was George, Duke of Buckingham, a notorious English rake protected by King Charles. But Parliament stepped in and prevented the king from protecting Georgie. Irate jealous husbands were about to attack him ("when the boys came out to play"). Then "Georgie Porgie ran away."

Jack and Jill refers to the beheading of Louis XVI and Marie Antoinette.

"Mary, Mary quite contrary" refers to Catholic Queen Mary, whose "silver bells and cockle shells" were instruments of torture used to make "her garden grow" with the corpses of Protestants. The "maids in a row" were guillotines.

The words of "Ring around the Rosie" are interpreted like this:

"Ring a ring 'a rosies" represents the raised red rash that was the first sign of black plague.

"Pocket full o' posies" refers to a pouch of sweet-smelling herbs carried around to mask the stench of death and protect from plague.

"Ashes, ashes" is really "atichoo, atichoo," referring to violent sneezing associated with plague.

"All fall down" refers to the fact that 60 percent of the victims died of black plague in England in 1655.[24]

THE VALUE OF A COLLEGE EDUCATION

It is not often that one of my students gets a job related to what I have taught him or her. Mostly they end up selling things. Perversely, this embarrassment has accrued to my advantage. One of my protégés sold me a rug—wholesale.

But one student achieved a modicum of success. He parlayed his knowledge of parasites into a job. He became a rat inspector. He was employed by the New York Port Authority to trap rats and look at their fleas, not just any rat flea, but the famous oriental rat flea, *Xenopsylla cheopis* (named after Egyptian pharaoh Cheops). The young man was eminently qualified for the position. He had studied that famous flea in my lab. How do you identify this rat flea from its cousin, the northern rat flea, *Nosopsyllus fasciatus*? It's easy. This common rat flea has a comb of bristles around its "neck."

But the human flea, *Pulex irritans*, is another story. If he found one on a rat, he needed to focus the microscope precisely because the human flea is almost identical to the oriental rat flea. Both lack combs. To differentiate he had to look at the first huge leg muscle (coxa). If the plasticlike shield covering the muscle has a diagonal pleat on it, it is *Xenopsylla*. If not, it is *Pulex*. Why the fuss? Why pay this kid big bucks to look at fleas all day? Because the oriental rat flea is the host of the bacterium *Yersinia* (= *Pasteurella*) *pestis*, which causes that most horrible of diseases, the black plague. The disease killed almost two-thirds of Europe's population in the fourteenth and seventeenth centuries. The plague made its last grisly visit in India between 1898 and 1908, when about half a million people died each year.

Lest you become complacent, there were 992 cases in the United States between 1900 and 1972—and 720 were fatal. Every year a variant of plague, called sylvatic or pneumonic plague and carried by ground squirrel and prairie dog fleas (and other West Coast rodents and their fleas), kills a few people, usually in California and New Mexico, evidence that fulminating plague is lying in wait. This makes the current tizzy over swine flu look like gossip at a tea party. If my student finds an oriental rat flea, the whole port of New York might be closed down.

In days of yore (fourteenth century and later), the black plague descended on Europe. Its cause was unknown. People invented all sorts of possibilities (smells, sneezes, bad air). One suspected cause was dead rats. Someone noticed that rats climbed out of their holes and died in great numbers when plague was about to descend on a town. Although the death of the rats was associated with the massive numbers of dead townspeople, removing and burning the remaining living rats did not solve the problem. The fact that the rat buriers died in droves seemed to substantiate the hypothesis. The problem seemed insoluble. The pandemic raged on despite rat removal. Town populations were very upset. Death was everywhere. Cartfuls of corpses were a daily sight. Something had to be done. In an emergency there was always one kind of suspect, the Jews. These perennial outsiders were herded into barns and burned alive . . . providing satisfaction if not survival. After the Jews were gone, the disease continued to ravage the population. Finally it "wente away." The prayers of the townspeople had been answered. No matter that they burned their Jews . . . God brought them salvation.

In reality, members of the population were selected for resistance to the disease—that is, spontaneous previous mutations made them less susceptible to the bacterium and they survived. This reservoir of resistant individuals numbered about 30–60 percent of the previous population. They formed a new "race" of resistant individuals. Henceforth this genetic grouping comprised the European population.*

The discovery of the relationship between dying rats and black plague—it is currently called bubonic plague—is a distortion of cause and effect. The effect was the dying rats—the cause was their fleas. That is, bacteria that the fleas carried. What actually caused the raging rate of infection of the townspeople?

The flea bites an infected person or rat, sucking in the infective bacteria with its blood meal. *Yersinia* reproduces rapidly in the dark, bloody digestive tract of the flea. A swollen plug of billions of bacteria forms, blocking ingested blood from entering the stomach. The sensation of an empty stomach makes the flea feel ravenous. Starving, it jumps from person to person in a futile attempt to assuage its perceived hunger, madly sucking blood all the while. It is these frenzied, frustrated fleas that are the villains. Thus the rats were only the intermediaries—they did not cause the disease. It was the flea-borne bacteria that were the ultimate etiological agents.

---

*Similarly, the European population consisted of survivors of diseases like smallpox and measles. When they colonized the New World, some had relatively harmless rashes—measles and chicken pox. These were harmless to them but destroyers of genetically unprotected Indian populations.

PLATE 22

A. ORIENTAL RAT FLEA, *Xenopsylla cheopis*. To 1/10 inch; dark brown; lacks combs; three pairs of legs. Diagonal pleat across largest leg segment, coxa, (closest to body)—not visible here—differentiates this flea from human flea, *Pulex irritans*. Male has coiled penis visible in abdomen.

B. FLEA FACES. Groups of large, dark spines on head, called combs, differentiate species. Those under eye are called genal combs; those resembling collar, pronotal combs.
> (1) Both combs: dog and cat flea, *Ctenocephalides canis* and *C. felis*.
> (2) Pronotal comb: northern (common) rat flea, *Nosopsyllus fasciatus*.
> (3) No comb: human flea, *Pulex irritans*; oriental rat flea, *Xenopsylla cheopis*.

C. ORIENTAL RAT FLEA is vector of plague bacterium, *Yersinia pestis*. Here, gut of flea is clogged with plug of rapidly reproducing bacteria, preventing food (blood) from reaching "stomach." Empty stomach creates sensation of hunger, causing flea to attack host after host, regurgitating infected blood as it feeds. Plug of bacteria visible in swollen gut.

D. BUBOE. Purplish, swollen, inflamed lymph nodes in armpit, groin; diagnostic of bubonic plague. Also purplish blotches under skin caused by subcutaneous hemhorraging that start with blackish pustule around bite. Two other *Yersinia* plague variations: pneumonic (lungs), septicemic (blood) do not or rarely produce buboes. Similar to buboes on neck characteristic of sleeping sickness (Winterbottom's sign).

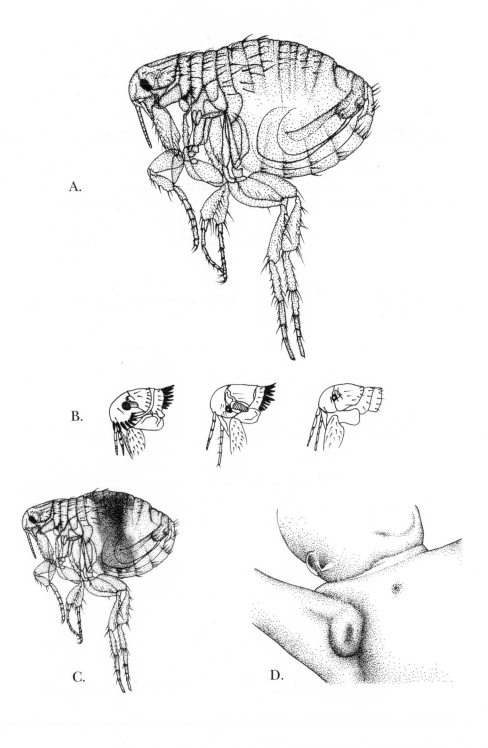

A.

B.

C.

D.

The symptoms of the disease began with the aforementioned rosy rash. Then came the buboes—egg-size blackish balls, enlarged lymph nodes under the arms and in the groin, grotesquely swollen chambers of deadly bacteria. Then the tips of the fingers and nose became black. The capillaries under the skin of the extremities burst, and the resulting region turned black from the metabolized blood (like a black eye). Unfettered bacterial reproduction permeated the vulnerable host's body with deadly toxins. His immune system was unable to cope. Sometimes the hideous buboes burst, their purulent population escaping in a cascade of pus. High fever accompanied the infection as the body attempted to kill the invaders by making their environment so inhospitable that they would die. To no avail; the invaders usually won.

After four or five days the afflicted often exhibited mental dullness, usually followed by hysteria. Then came a coma from which they did not revive.

23

HOW TO GET RID OF CRABS

New Orleans, Tulane University parasitology laboratory, 1963. A blushing fraternity boy deposits a vial on my lab table and runs out. At first the contents are not discernible. I hold the vial up to the light. In it are about ten coarse hairs. Little white dots move among the hairs. What are they? A creepy sensation seeps into my psyche. *These are pubic hairs and the dots are crab lice!* College fraternities are brotherhoods of experimenters. This frat brother had gone to Bourbon Street to experiment and had gotten into trouble.

A similar event occurred at the office twenty years later. Betty, the department secretary, complained that every winter her husband behaved badly. He was restless and cruel.* She was happy when he went back to Sicily to visit his sainted mother. Each spring he returned. Conjugal life resumed. Their pent-up longing was relieved. One spring day she came to my office. I recognized that embarrassed look: red-faced, she handed me THE VIAL. I took one look at it and realized that it contained the characteristic spotted pubic hairs. Sheepishly, she asked, "What are these?" I had to explain to her that like venereal diseases, rarely does one pick up crab lice from the toilet seat, her husband's protestations notwithstanding. After a few more melodramatic winters they were divorced.

---

*It is likely that he was afflicted with a condition known as seasonal affective disorder (SAD), characterized by depression and aggressive behavior.

A LOUSY STORY

Three species of lice live on humans: head and body lice that can be distinguished by their habitat, and crab lice. Head lice (cooties) live in fine hair. Their eggs are called "nits," and they spread rapidly from child to child in elementary school. When they are diagnosed, mom has fits. She boils everything—sheets, pillows, curtains, clothing—she even looks fixedly at junior.

Body lice live on the somewhat coarser chest and underarm hair, but there may be other criteria, such as temperature, for their choice of habitat. The species are able to interbreed and their appearance is virtually identical, leading modern scientists to suggest that they are subspecies. Confusing the taxonomic determination are minor differences: the slightly smaller head louse, *Pediculus humanus capitis*, produces 80 to 100 eggs at a time, while the body louse, *P. humanus corporis*, produces between 200 and 300 eggs. Are these differences sufficient to consider each louse a separate subspecies? Nowadays, because their differences are so inconsequential, some consider them different races.

It seems likely that the body louse evolved from the head louse when clothes were invented. In cold countries body lice are as prevalent as head lice; in tropical countries, where less clothing is worn, body lice are uncommon. In fact, body lice often live on clothing almost exclusively, returning to the body to feed. Medical texts have been known to recommend that an infestation of body lice can be eliminated by not wearing clothes for three or four days (not recommended for cold climates).

THE DISADVANTAGE OF WEARING CLOTHES

Recent DNA research has pinpointed the moment when body lice sprung from head lice. This seems to have occurred when humans invented clothing, between 72,000 and 42,000 years ago.* But there

---

*DNA of head lice has been dated to thirteen thousand years ago in the New World. Dried lice on decapitated, mummified Peruvian Indian heads were recently precisely dated at AD 1025, proving that they were in the Western Hemisphere before Columbus

is a disadvantage to wearing clothes that must have had evolutionary significance. Nakedness reveals all—including whether or not the potential mate has lice. Is it possible that deep in the residual memory of the modern woman is the instinct to reveal as much skin as possible to assure her suitors that she does not have lice? Is it possible that one of the reasons that clothes were invented was to hide the fact that the potential mate had lice?

In the past, when baths were a luxury only for the wealthy, lice were the ever-present companions of humans. Even the rich and anointed were not immune. The story is told that the morning after King Henry VIII caused the assassination of Thomas Becket, archbishop of Canterbury, his attendants removed the vestments from the cold corpse—and out boiled a horde of lice.

Because lice were an ever-present source of irritation, the term "louse" was coined as a pejorative for an unpleasant person. An individual who is "lousy" suggests that he or she is infested with lice. The term "nit-picking" refers to going over a person "with a fine-toothed comb," to remove louse eggs. Other references to nit-picking suggest excessive precision or reducing something to its essentials as "getting down to the nitty-gritty." "Nitwit" does not need interpretation.[25]

Although lice have a negative connotation in humans, they serve a more congenial purpose in other primates. The social castes of baboons are delineated by who is picking lice off whom. The alpha male is attended by the alpha female. She is deloused by her subordinate, and so on. Most monkeys spend the major part of each day delousing each other, confirming the hierarchy.

HAIRY LAIR

Lice are insects. They lack wings and are so flat that it is difficult to scratch them off. Each one must be purposely plucked from its almost invisible hairy lair. Six legs terminate in rigid, clawlike "fingers" that

---

and were not another curse to be added to the bad reputation of the Spanish Conquistadores. But they may have transferred louse-borne typhus from the Americas to Europe.

hook onto individual hairs. They copulate on the host's body. The male crawls under the larger female from behind. When the tips of their abdomens fuse, the female assumes a vertical position, lifting the male. They then settle down horizontally and remain attached for about thirty minutes.

During feeding, stiletto-like proboscis parts make an incision in the skin. So thin and sharp are they that the wound is painless. Then other mouthparts are inserted, probing sensitively for a capillary in the dermis. Once the blood flows, the muscular pharynx contracts to create suction and blood is sucked up the proboscis and into the gut. In a reverse flow, the insect's saliva pours into the host's blood. If it were not for the saliva, the blood would begin to clot after a few seconds, interrupting the meal before the louse has enough nutrients to support its future brood. The copious saliva contains an anticoagulant that prevents clotting, providing time for a substantial meal.

It is the proteinaceous saliva that causes our misery. By challenging the immune system, the saliva causes a localized immune response characterized by enlarged capillaries. The swollen capillaries become a red, itchy, or painful bump on the skin. *It is not the stabbing of the skin that is so painful, but the subsequent immune response.*

Human skin subjected to louse bites over a long period may become darkly pigmented, a condition known as vagabond's disease. People with this affliction often become immune to further irritation.

Lice transmit a variety of diseases: relapsing fever, typhus, and tsutsugamushi. These are the diseases of the unwashed or crowded. Trench fever (a form of relapsing fever) infected many soldiers in World War I, and typhus was a dreaded disease until bathing and clean underwear became common. During the Vietnam War another of the innumerable unrecorded tragedies was an outbreak of relapsing fever.

Louse-borne diseases are introduced via a primitive transmission method, posterior station, referring to the source of infection, the insect's feces.

The most dangerous louse-borne disease is typhus. Symptoms begin with a very high fever that lasts about two weeks, succeeded by aches and pains, a rash, and sweating. It is often fatal. Relapsing fevers are debilitating but rarely cause death.

HOW TO GET RID OF CRABS

Crab lice live only on coarse (pubic) hair. They are so-called because each of their six legs terminates with a huge immovable claw, the better to grasp the pubic hair. This louse is rarely active on the host, keeping its mouthparts buried in the skin for long periods, sucking blood all the while. When stimulated, that is, participating in the antics of a copulating human couple, it moves with uncharacteristic rapidity and agility to the new host. No wonder. Since each female can lay thirty eggs in her lifetime, the pubic region can become overcrowded. The new host's uninfected pubic region can be "like a breath of fresh air" to a crab louse escaping from a crawling mass of comrades.

A crab louse infestation results in fierce itching, leading to embarrassing moments when the persistent itch cannot be relieved unless by scratching. The realization that these "bugs" are crawling around in one's private parts is more than one can bear. I am often asked, "How does somebody—not me—get rid of them?" Here is a time-honored recipe:

1. Shave a one-inch strip through the middle of the pubic hair.
2. Rub kerosene onto one side.
3. Light it.
4. When the lice run across the strip, stab them with an ice pick.

Every year in parasitology class, I sponsor a contest to determine who is most creative among the students. The objective is to make up a common name for crab lice. As a source of inspiration, I begin the assignment with examples of wise wife's creativity: "bloomer bunnies," "crotch crickets," "vulva vermin." These alliterative nicknames seem to present the ultimate challenge. Even the sweet young thing in the first row contributes a list of horrendously vulgar creations. Below is a distillation of forty years of contributions (list cleaned up for public consumption):

PLATE 23

A. CRAB LOUSE, BLOOMER BUNNY, *Pthirius pubis.* To 1/16 inch; six legs; body wide; white dots on pubic hair usually eggs (nits). Adults often on skin, feeding; also on armpits, beard, moustache; gray-blue color on skin at feeding site. Insert mouthparts directly into blood vessels. Crablike claws on second and third pairs of legs; dots along sides of abdomen are "nostrils" (spiracles). Next time you have a severe itch in your pubic area, be aware that there is a strong association between crab lice and STDs.

B. INFAMOUS VIAL delivered by "innocent victims." Pubic hairs with nits; some live lice were seen as moving dots.

C. BODY/HEAD LOUSE, *Pediculus humanus corporis, P h capitis.* To 1/16 inch; six legs, body narrow. Eggs are white dots cemented to head hair, clothing (*P. corporis*), called nits, often on back of neck or behind ears. Female lays nine or ten eggs per day; three hundred in her lifetime; hatching time determined by external temperature. Bite causes red papule, intense itching. Body lice will die if clothing (underwear) changed daily. Head lice nits often detected by school nurse inspection.

D. CLAW OF CRAB LOUSE. Photomicrograph shows double "fingers" of second and third legs; adapted to grasp pubic hairs.

E. NIT OF CRAB LOUSE. Photomicrograph shows attachment to hair shaft (*above*) and breathing holes in lid (operculum).

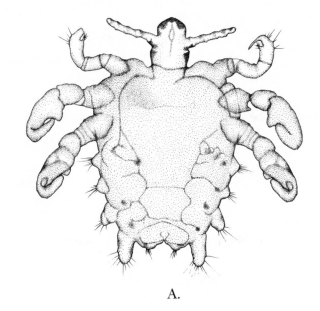

A.

C.

B.

D.

E.

Pecker wrecker
Loin leeches and labia leeches
Groin grazers and crotch cooties
Bidet beasts
Coitus preventus
Vulva varmints
Testicle ticks and nut nits
Bush whackers
Pornographic pests
Jock jackals and jock joggers
Ribald raptors
Testicular terrorists
Streetwalker specials and call-girl contributions
Nut newts
Vagina vipers
Mound munchers
Peter eater
Teste ticklers
Testosterone trophies
Vaginal vampires
Fur fairies and pelt plankton
Labia locust
Pudendum pariah
Gonad gophers and love lobsters
Harlot's honeys and tart's treasures
Claws in your drawers
Vagina vexers
Pubic picnickers and booty cootie
Sexy scamps
Rug rats

24

WILD VIRGINS

THE B-17 FLYING FORTRESS OF WORLD WAR II vintage looked ominous. Undefeated by war and years of service, it roared aggressively sitting on the runway. Engines revving, it took on the last of its cargo through open bomb bay doors. With a shudder and rattle, the plane took off in a cloud of dust. Its mission was to cross the border and to drop its "bombs" over a densely populated area. Was this a mission during World War II? No, the targeted population was—cows. Herds of cattle peacefully munching on rich Texas grass and, across the Rio Grande, on the prairies of nearby northern Mexico.

The real enemy was a fly. The screwworm fly, *Cochliomyia hominivorax*, plagues cattle, making them nervous, thin, and, in ranching parlance, "unthrifty," causing the loss of millions of dollars to the cattle industry. Thus the declaration of war against the fly. The hope was not to eliminate this pest, for the ranchers realized that such an undertaking would be impossible, but to reduce its numbers so that its depredations would lessen their economic burden. How to attempt such a task is a story of desperation giving rise to ingenuity. Visualize a bunch of cowboys sitting in a saloon discussing their losses from fly infestations. One of them, probably a former World War II pilot, knew enough of the fly's life cycle to come up with an outlandish plan— bomb them into submission.

The screwworm fly is unusual among insects. It is not promiscuous. That is, only one male can father the hundreds of eggs produced by the female. It has been suggested that the male produces a substance that remains on the surface of the female's genitalia that causes her to "not be in the mood." Once inseminated, she retains the sperms for future use.

Eggs are laid when the female senses a suppurating wound. This phase of the life cycle is almost ensured, for cattle, infested with ticks, seek relief by rubbing themselves against any object that is at back level. On the treeless prairie, barbed wire fences beckon. Almost every steer is covered with bleeding wounds on its back.

Drawn by the stench of a pus-and-lymph-weeping wound, the female lays up to a hundred eggs. They hatch into distinctive white, spiny fly larvae called maggots. Their breathing openings, spiracles, are located at the posterior end of the body. The maggots "screw" themselves into the injured tissues and eat their way through the cow's body. They migrate under the skin in a manner that would make a sci-fi horror film look tame, devouring tissues in their subcutaneous meanderings. To breathe, they digest a hole in the skin of the cow, projecting their spiracles out of the hole. This causes such damage that the skin loses value as a prime source of leather. The irritation caused by these internal migrations and the lesions caused by the breathing holes distract the cattle, making them edgy, and they do not feed properly. They can die of a secondary infection. Eventually, the sated maggots mature, climb out of a hole drilled into the host from inside out, and fall to the ground, where they find a "prairie pancake." Burying themselves in the cow dung, they pupate, to eventually emerge as adult flies, continuing the cycle.

The bombs used against these screwworm flies did not contain explosives. They were paper bags filled with male screwworm flies. Upon landing, the bags burst open, releasing a horny horde. These males had previously been bombarded with gamma rays, making them sterile. Once they mated with a female, their damaged sperms rendered the production of viable eggs impossible. The wild virgins were fooled.

Once fertilized, they were trapped by their fidelity and could not mate with a fertile male.

However, the effect of this early effort at biological control was not as profound as expected. It was discovered that macho Mexican males were meandering across the border and mating with the virgin females, keeping the fly population higher than desired. Complicated negotiations were entered into with the Mexican government. The Mexicans were not enthusiastic about having bombers cross their border. Once permission was obtained, drastic declines in the screwworm population occurred.*

HUMAN EATER

Hominovorax means "human eater." *Cochliomyia hominovorax,* the New World screwworm, was so-named by Charles Bequerel in 1858 because he attributed the deaths of hundreds of prisoners in Devil's Island Penal Colony of French Guiana to this fly. Its range is limited to warm regions. However, migrations can occur from Texas and Mexico northward in the summer, even as far as Kansas.

In 1883 a Dr. Richardson reported a case of human infestation by *Cochliomyia hominovorax* in the *Medical Monthly* of Peoria, Illinois. A traveler in Kansas was sleeping when a fly laid its eggs in his nose. The fly was probably attracted by a discharge of mucus. The first symptoms were those of a severe cold. As the maggots cut away through the tissues of the head, the patient became slightly delirious and complained about the intense misery and annoyance in his nose and head. The maggots finally cut through the soft palate, impairing his speech, and then invaded the eustachian tubes. Despite the removal of more than 250 maggots, the patient eventually died.[26]

In 1988 the New World screwworm (as it is called to differentiate it from a European species) turned up in Libya, to much consternation. If it could travel oceans, why could it not cross the Sahara

---

*The sterile irradiated male technique is now being tried against medflies and tsetse flies.

and threaten the whole continent of Africa, which has its own fly problems and doesn't need ours? Within a year over two hundred cases of human myiasis (fly infestations) were reported around Benghazi.[27]

FLORENCE NIGHTINGALE

The Crimean War (1854–56) took a massive toll of lives. As in the subsequent U.S. Civil War, more deaths came from disease than from bullets. A tour through the miserable tents that were hospitals was to walk through unfathomable horror. Blood, flies, and excrement were everywhere. To be hospitalized was virtually a death sentence. Seeing this, an unusually competent and compassionate woman named Florence Nightingale set an example by rolling up her sleeves and wading into this morass of terror. Her noble example became the foundation of the nursing profession.

During those same battles, the physicians noticed that injured men who were secondarily infected by flies had a higher survival rate. Careful examination revealed that only one species of fly was of benefit, the large metallic blue blowfly, aptly named *Calliphora vomitoria* (possibly the "blue-tailed fly" of the folk song). This fly senses the stench from an untended, open wound. It lays twenty or thirty eggs on the periphery of the pus-covered lesion. The eggs hatch into maggots. They eat only the decaying tissues, leaving a perfectly neat area of healthy tissue capable of repairing the wound unimpeded by rotting flesh. The maggots cannot penetrate healthy tissue.

During the U.S. Civil War the lessons learned by the Crimean War surgeons were put to use. Blowfly maggots were deliberately placed in open wounds to cleanse them of damaged tissue, making recovery more certain and rapid. This practice was used for years after World War I, until the discovery of sulfa drugs and penicillin. Modern science has discovered the basis of this historically important fly-human interaction. The saliva of the maggots contains a bacterium, *Proteus mirabilis*, that produces substances that kill pathogenic bacteria and promote wound healing.

THE SLEEP OF DEATH

Blood is the all-providing source of survival of the dreaded tsetse fly, *Glossina*. It bites a sick person, picking up elongated one-celled flagellates in the blood. Soon the fly's body is literally filled with the slender infective stage. They spill over into the fly's huge salivary glands, huge because they must produce a flood of anticoagulant-rich saliva to prevent the blood from clotting as the fly sucks up its rich red meal. In the saliva are agents of death, trypanosomes.

The fly bites again, injecting the parasites into a new human host. Once in the blood, they go through a series of rapid cell divisions and massively increase in number. In days, vast millions flood the blood, their presence causing a central nervous system disease called sleeping sickness. What accounts for the sleepy symptom of this dreadful disease is not fully understood. But the most dangerous form of the disease kills its host too rapidly for this symptom to appear. The causative agent, *Trypanosoma brucei rhodesiense*, swims in the blood plasma and has rarely been found in the brain. Something mysterious is happening.

Part of the mystery is the modus operandi of the parasite and its unique method of resisting the body's defenses. This chameleon of parasites can change its identity with a thousand genes to use to confuse the immune system.* No sooner has your body produced one antibody than the parasite changes its identity and another antibody has to be produced. The immune system goes berserk! This overwhelming array of antibodies may cause an autoimmune response. One of the many antibody types evoked by the ever-changing trypanosomes may latch onto the host's brain nerve cell sheaths.

Or rampant antibodies may induce an autoimmune reaction whereby red blood cells are not discriminated from the parasites and

---

*Recent research has uncovered the mechanism for parasitic protozoans to change their immunological identity. The ability to do this is written on a thousand genes. It may be possible to force the parasites to express *all* their "multicolored coat" proteins at the same time. In that case, any antibody will be effective. This research was done on *Giardia lamblia*, a common human parasite discussed in the last chapter. It remains to be seen whether or not this theoretical cure will be effective with trypanosomes.

are destroyed, causing massive anemia, depriving the brain of oxygen. At this point, no one answer provides a full explanation.

In contrast to the quick death of Rhodesian sleeping sickness, the trypanosome that causes "real" sleeping sickness, *T.b. gambiense,* kills slowly. Over many months the host is "sleepy" all the time, to the point of being unable to function—even to secure food. This species has been found in the brain. But the pathology of this chronic sleeping sickness syndrome has not yet been fully explained.

The disease occurs in central Africa in the vicinity of streams and is often called riverine sleeping sickness because its tsetse fly vector, *Glossina palpalis,* lays its eggs on the undersides of leaves of plants that grow along rivers and streams.

The human host dies of causes related to long-term stupefaction, including starvation and secondary infections. Before the lassitude, the following symptoms have been reported: "When the parasites enter the brain, victims hallucinate wildly. They have been known to chase neighbors with machetes, throw themselves into latrines and scream with pain at the touch of water."[28]

An early symptom of Rhodesian sleeping sickness is bulging, swollen lymph nodes on the back of the neck called buboes.* (Does the term "boo-boo" originate from this condition?) The swollen buboes are a death sentence. To find them is to know that the host is doomed to an early death, possibly within weeks. Captains of slave ships would bring their human cargo on deck to examine their necks. Anyone with the characteristic swellings would be summarily thrown overboard. No use feeding these doomed individuals, they would never survive the trip to America.

TOUGH OLD BROAD

Professor Priscilla P. was a charming, laid-back woman in her sixties when I took her course. Her deceptive indifference to minutiae was an

---

*Buboes on the neck are now called "Winterbottom's sign" after the British officer who first described the symptoms.

inspiration. Her lab on crustacea began when she sent out someone to a Chinese restaurant to bring back lobster, Cantonese-style. Before eating, each of us followed her directions for dissection of a lobster, vaguely uncomfortable with the knowledge that we were about to eat similar specimens.

Her philosophy profoundly affected my own teaching style.

Years later I had occasion to invite her to a workshop. In the pre-meeting gabfest, she told this story: Once, at a refresher course in parasitology at a university in New Orleans, her lab partner was about to inject an experimental rabbit with a heavy dose of *T.b. rhodesiense.* A telephone call. Someone brought a phone. She turned to take the call and the hypodermic needle accidentally penetrated her arm. *She was injected with the virulent strain of sleeping sickness.*

Soon her lymph nodes began to swell with incipient buboes.

New Orleans was famous for being a center of research in parasitology. It would be embarrassing to the staff of world experts if the only case of death by sleeping sickness in the United States occurred there. She was admitted to hospital. Eminent parasitologists walked around her bed wringing their hands. What to do? Finally someone thought of a new medicine that was being studied in England. Its mode of action was to raise the body temperature to unnatural heights, theoretically killing the microscopic parasites before the relatively voluminous host. The drug was rushed to the scene and administered, inducing a fever of 106 degrees Fahrenheit. The parasites died and the tough old broad survived.

In 1980 new drug, eflornithine, was discovered. Its success led to the name "resurrection drug" because it was able to pull the dying out of comas. It was unprofitable because most of the victims were impoverished and couldn't pay for the expensive pills.

By late 2000 the pills were running out. Then a topical form emerged. It was the key ingredient in Vaniqa, a cream to prevent facial hair in women. "After critics accused [the company] of catering to vain rich women while letting poor African women die, the company agreed to make an injectable form . . . and now gives it free to the World Health Organization and Doctors Without Borders."[28]

PLATE 24

A. NEW WORLD (PRIMARY) SCREWWORM FLY, *Cochliomyia hominivorax*. To 5/8 inch, bluish gray; eyes red; three thick lines on thorax; broad gray band on hairy abdomen.

Larvae cannot penetrate skin; must enter through suppurating wound; monogamous; Florida fly factory, in 1958–59, released 2.75 billion sterile males; reduced screwworm population from forty thousand a year to 0.

B. SCREWWORM LARVA EMERGING FROM SKIN using fanglike mouthparts at anterior to cut through flesh. When immature, posterior is uppermost with crescent-shaped nostrils (spiracles) poking out hole at tip of "mosquito-bite" to breathe (cf. pl. 30). That's right, nostrils are near anus. After a few weeks feeding inside cow, larvae reverse position, escape by drilling holes through cow's skin (making leather less valuable); fall to ground, pupate; emerge as adults. White body has rings of black spines on ridges, making larvae difficult to remove. Bot fly larvae behave like this in people (plate 30).

C. BUBOES ON NECK OF AFRICAN SLAVE are symptom of Rhodesian sleeping sickness caused by bite of tsetse fly, *Glossina morsitans*, that transfers *Trypanosoma brucei rhodesiense*. Buboes are swollen lymph nodes, to 1½ inches.

D. SLEEPING SICKNESS PROTOZOAN, *Trypanosoma b. rhodesiense*. Three times the diameter of red blood cells; single flagellum along outer edge of transparent undulating membrane, is swimming in watery plasma surrounded by donut-shaped (biconcave) red blood cells (dark spaces in rbcs are depressions, not nuclei).

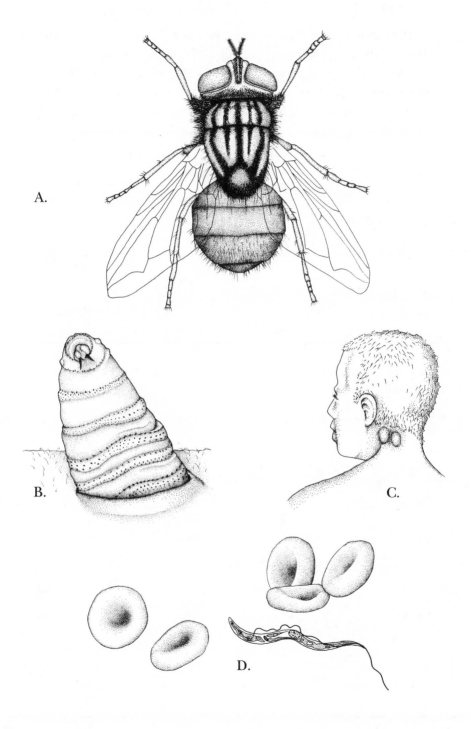

A.

B.

C.

D.

The meeting adjourned late, and the last train was soon to depart. After a hair-raising midnight car ride to the station, we heard in the distance "All aboard!" What to do? Priscilla was marooned. Not to worry. The tough old broad darted down the platform and leapt onto the train, disappearing into the night.

INEXPLICABLE BEHAVIOR

SOME RELATIONSHIPS ARE MORE INTIMATE THAN OTHERS

RELATIONSHIPS ARE AS UNIQUE AS THE PARTNERS. Some are short-lived; some so intense that both partners can't live without one another. Sometimes each partner has several relationships over time.

Why is there such variability, so many unique associations? How have these complex interactions evolved? Symbionts, in their dependent relationships, can't realize that they are "helping" each other to survive—or causing harm to the host. Nothing in the nonmammalian world can be ascribed to benevolence or malevolence. Purely functional, each behavior has evolved over eons to become a pinnacle balanced on precursors—a pyramid of ever-closer ancestral intimacies aimed at the survival of the symbiotic species. Even advanced animal behavior can be described in mechanistic terms: your dog's affectionate behavior is merely a human-selected regression to puppy levels. Dogs are bred for infantilism.[29] The perpetual puppy endears itself to its owner using the behaviors it has evolved to stimulate its parent to feed and care for it. This hard-hearted view may seem contradictory to your experience. Maybe dogs are capable of affection and this is an exaggeration?

But what about birds? The mother bird flutters to draw the predator away from her nest. Is this apparently clever ruse a brilliant way of preserving her genes in her offspring? No, her flopping around on the ground is more likely the result of a conflict between her fear behavior (the urge to fly away) and her instinct to protect her young.

The female alligator charges an intruder threatening the pile of rotting organic matter containing her eggs. When she hears the piping voices of her young in the rot, she digs them out and, with huge, flesh-tearing teeth, gently carries them to the water. Is this maternal devotion? Is it learned, variable behavior—or is it inborn, emotionless? Is there emotion in the baleful eye of the alligator?

What birdy mother and her slithery cohort have in common with the mammals is that they are all vertebrates. But the emotional line seems to stop at the mammals.

BEHAVIORAL INTERACTIONS

What about other-than-parasitic intimate behavior between two animals of different species? The parasite-host interactions described in this book usually incorporate physiological adaptations.

Mutualistic and commensalistic relationships sometimes appear to lack a physiological component, but there can be depth to the interaction. Sometimes intimate *behavioral* dependencies do not lead to the traditional feed-and-be-fed interactions such as occur between parasites and their hosts. Sometimes, like the "human-eater" flies, part of the cycle is parasitic and part free-living. But these parasites *require* a mammalian host. Occasionally symbionts can live independent lives; sometimes they are obliged to live in or on the host to survive.

It is the "stand-off" noncontact behavioral interaction that fascinates. Apparently some species have "learned" to rely on each other in all kinds of closely interwoven behaviors. Their genes, capable only of synthesizing a variety of enzymes, should not be capable of catalyzing behavioral intimacies. Surprisingly, the genes, those bits of DNA, become involved in complex behaviors requiring an interplay between two animals, like two dancers twirling around each other in a complex pas de deux. She pirouettes, stimulating her partner to hold her while she twirls on toe; then her fluttering arms signal him to lift her off the ground, showing off his grace and form. They move offstage, entwined. This complicated interaction has been practiced a hundred times. Both dancers have spent many hours strengthening themselves for this choreographed series of movements.

But suppose that this rendition of *Swan Lake* was genetically dictated and they were born with the steps and capabilities and could perform them from birth?

The next few chapters will describe examples of the continuum of the infinitely complex choreography of nature, the dances of the genes, the pas de deux of the clownfish and its partner sea anemone; the cleaner fish flirting with the host parrotfish, and other intimate behavioral interactions.

25

TOPSY-TURVY WORLDS

BRIGHT GREEN HEMISPHERES three feet across are jumbled together with craggy hillocks, columns, flat bumpy plates, and stony fingers, all crowding the sunlit shallows of the coral reef. At the crest of the reef, orange elkhorn corals, *Acropora palmata*, break the force of the wind-driven ocean waves into white, bubbly foam and spray. The shockingly bright sunlight illuminates a gorgeous, colorful panorama. Soon the shallows slope into wide plains of orange-green staghorns and ten-foot-high bright green hillocks of mountain coral covered with multi-colored sponges.[30]

The scene dims as shadows of clouds dull the bright colors, but the fluffy cotton puffs soon move on and the sun returns to reveal sparks of every possible hue that flash momentarily as reef fishes seek havens under coral ledges and crannies. Amid this kaleidoscope of color, you don't notice that the predominant color of the reef is greenish, varying shades from tan to golden, but greenish nevertheless. Bright yellow and purple sea fans and other soft corals deceive. The reef is a riot of color in your mind. But inside the transparent tissues of the soft-bodied polyps that make the coral reefs are golden green one-celled algae.

You move seaward. The water deepens. The exquisite color disappears with the sunlight as the bottom slopes into the distant mysterious darkness of the black abyss.

THE BLACK ABYSS

Imagine a huge dark planet. No light. No oxygen. Giant fissures slash the oozing surface. From their black-on-black interiors spurts a noxious acidic fluid, so awful in its smoking, boiling, sulfurous fumes as to re-create a freezing Dante's inferno. This is a black, dead hell, as terrifying in its icy darkness as Dante's flames. Life as we know it cannot exist here in this unremitting frozen crushing darkness.

A mile below the blindingly sunlit coral reefs that cap the continental shelf, there is a flicker of movement. The beams of a submersible's lamps cut through the thick darkness like flaming arrows, revealing a vast desert of ooze. The bottom is pristine, undisturbed for millions of years until the submersible's propellers roil the ancient velvety surface. This fine sediment was formed by a constant rainfall of an almost infinite number of tests (shells) of microscopic amebas and glassy one-celled diatoms falling from the sunlit photic zone.* The fine ooze bottom rolls on formlessly into the darkness beyond the penumbra of the submersible's light.

The viewer's senses become dulled, lulled by the monotony of peering through the submersible's window at the constant dun-colored, undulating flatness of the bottom. Suddenly, the featureless abyssal plain is punctured by a huge dark cone spouting black "smoke" that rises from inside the Earth's mantle as if from the depths of hell. This miniature volcano is vomiting up stinking gases that have been churning through the hot, rocky cap that surrounds the molten core of the Earth. The ocean's water has found fissures in the rocks and it flows underground, heated to extreme temperatures but not boiling, until released through a hole or crack. Gradually, when this mineral-laden, heated subsurface flow meets icy abyssal water, minerals precipitate out to form spectacular "black smokers."

As the submersible approaches the startling scene, a vague underwater panorama looms ahead. Dark shadows become black cracks

---

*The photic zone is the region of the ocean to which light penetrates enough to sustain photosynthesis. It varies widely with the transparency of the water. In general, the more plankton, the shallower the zone.

in the bottom called vents. They too spew super-heated water that burbles upward with force—freely flowing columns of water released from the pressure of constricting rock fissures.

The viewer peers through the hot-water columns and shakes his head in disbelief at the scene surrounding the vents. What cannot exist, exists. Indistinctly visible through the density-distorted, wavy water is what appears to be an unearthly colony of bizarre, unfamiliar animals. The submersible penetrates closer, moving through the blurry water until density differences dispel and clarity returns. Here, a mile and eons down, is a crowded, previously inaccessible world. Clusters of alien life forms surround the hot vents and cones. The most foreign of these clusters are thickets of inch-thick spectacular worms extending from white tubes. Some project six feet from the bottom. Their blood-red, cylindrical bodies glow in the shockingly bright light. Huge clams and mussels lie in random patches nearby. All seems still until starkly white crabs scuttle from their hiding places, their hysteria triggered by the never-encountered beams of dazzling light. But the crabs look different; so do the clams. The differences are vaguely disquieting, but the huge, bloody-looking, wormlike animals resemble no other earthly organisms. This is an alien community, no different in its strangeness from imaginary invaders from another world. Perpetual blackness and crushing depths have sheltered from the human eye an aggregation of animals unthinkably different from surface communities.

No plants can exist in this perpetual darkness. But plants sustain all life as we know it. No surface community can exist without plants. Their photosynthetic energy-trapping processes create the bottom layer of the trophic pyramid: herbivores eat plants, carnivores eat herbivores. Here in this dark world there are no plants—therefore no herbivores, therefore no carnivores.

The sun, the ultimate provider of energy, does not shine in these black depths. This life, this crowded, diverse community, defies everything we know about survival on the surface. How can this be? If we found such a community on Mars, it would be equally inexplicable. This really is a colony of aliens.

DESERT DWELLERS

Coral reefs are as inexplicable as these abyssal vent communities. The clear tropical seas hardly filter the sunlight, maximizing the solar energy that powers photosynthesis. But where are the plants? There ought to be plants. It appears that the coral reef trophic community is upside-down. Vast biomass of coral animals at the top, and at the bottom, the least visible—plants. Unlike the abyssal animal vent community, photosynthesis is possible, but the apparent paucity of plants seems inadequate to support the vast number of animals on coral reefs. To make this conundrum even more confusing, clear tropical seas lack the minerals that are needed for plants to flourish.

To the excited observer, the coral reef teems with wonderful animals. No matter that plants are not overwhelmingly evident. But to the biologist this is impossible—plants are the ultimate source of living energy. Here we face another biological impossibility.

The mystery has been resolved. The rocklike mounds, staghorns, brains, and fingers of coral have one thing in common: all are shades of golden green (or the green can be masked by superficial colors, like tan or orange). What makes them green? The answer to this question is the crucial clue.

Each coral colony is covered with a thin sheet of animal tissue from which project, like living hairs, tiny, flowerlike polyps whose "petals" constantly grasp at the few microscopic, animal-like zooplankters that swirl by. Among these dots are one-celled dinoflagellates. Some bridge the gap between plants and animals. They swim with two whiplike flagellae but contain chlorophyll, the one substance that can convert the sun's energy into food and oxygen. Sometime in the distant past, dinoflagellates called zooxanthellae found a home inside coral tissue in an extraordinary mutualistic relationship. The coral animals provide carbon dioxide and water from their metabolism to their internal "plants," the zooxanthellae. They, in turn, use the coral animal's metabolic products, carbon dioxide and water, to make food and oxygen through photosynthesis in the fundamental feedback system of interdependence between plants and animals. This relationship is so tightly knit that when cultured in a dish and coral tissue is added, these zooxanthellae produce a burst of food. They are "feeding" their hosts. But the ultimate energy source for photosynthesis is the intense tropical sunlight.

Coral polyps "farm" greenish dinoflagellates in their transparent tissues. That's why all hard corals are, underneath, shades of golden green. That's how the reef can survive in the almost-sterile, aquatic deserts of tropical seas. These "plants" within coral tissues take on the role of terrestrial trees and grass. The corals are the cows; the zooxanthellae, the grass. This mutualistic miracle sustains the vast explosion of life found in the nooks and crannies of the reef, all energized by the sun.

LOOKING AT THE PRIMEVAL EARTH

In the beginning God created the heavens and the earth.
Now the earth was a formless void, there was darkness
over the deep, and God's spirit hovered over the water.
—GENESIS 1:1

In the seemingly God-forsaken deepest depths of the ocean we see, as if frozen in time, the origin of the Earth—the darkness, the deep, formless depths. The origin of all life is here too, bringing to our consciousness a primordial world without oxygen. Truly, we can step back in time to the period before life as we know it appeared. No oxygen existed because chlorophyll did not yet exist. No chlorophyll, no photosynthesis—and no oxygen to support metabolism as we know it. But in this dark ancient world a primitive kind of metabolism appeared, one that did not depend on oxygen, one that might exist today on another planet.

On our planet, we envision the chemical evolution of life starting without oxygen. In the beginning, photosynthesis was impossible. Theoretically, eons after the first billion years of the Earth's formation, after gigantic storms of toxic methane and sulfides had abated, the Earth was ready. Nearer the sun than time-frozen Jupiter and Saturn, liquid methane warmed to form a gaseous, hydrogen-rich atmosphere. At first the atmosphere was a mixture of gases dominated by methane and sulfurous fumes. Then water appeared in the next billion years. Photosynthetic pigments evolved. Oxygen inevitably followed, replacing the toxic atmosphere.

But first the methane-rich atmosphere gave birth to the precursors of life. At some point in time, long strands of attached carbon molecules were driven together by energy provided by some source like lightning or then-rampant ultraviolet radiation. These precursors of life were able to absorb then-abundant, energy-rich molecules of sulfur compounds. The earliest metabolism appeared, extracting energy from hydrogen sulfide.

But sulfur-based metabolism is inefficient. It cannot compete with the oxygen-based metabolism that replaced it. Hence, in today's virtually oxygen-free abyssal bottom, ancient metabolic processes still exist. Frozen in time, the unchanged ancient Earth lies before us. Billions of years of evolution have not happened.

How can the huge, unfamiliar, blood-red worms function in the absence of an oxygen-based metabolic community? They can't. Nor can they extract energy from sulfur. Only an organism whose metabolism has not progressed beyond its ancient origin can use this primitive process. Bacteria qualify. We know that sulfur-metabolizing bacteria exist because in the anoxic mud of modern marshes, a spadeful of sediment smells horribly of rotten eggs—hydrogen sulfide—a sure sign of sulfur-based metabolism based on sulfur-metabolizing bacteria. Oxygen does not penetrate through the fine mud, and competition with the oxygen-using world does not exist.

In the inky depths, heat is injected into the abyssal community by hot water coming from the fiery depths of the Earth. The vent community is trapped by its need for warmth. The crowded population surrounding the vents can extend out only as far as warmth can sustain the primitive bacterial metabolism.

But what has all this to do with the survival of the vent community? We have seen that sulfur-based metabolism is confined to bacteria and only bacteria. How can the metabolic demands of the worms be satisfied? Samples drawn from the deep have answered this fundamental question.

The abyssal vent organisms "farm" the sulfur-metabolizing bacteria in their tissues to extract energy in a mutualistic relationship similar to that of the zooxanthellae- dependent corals a mile above. The similarity extends beyond the obvious. New information reveals that the bacteria use carbon dioxide dissolved in the water as a source of energy to break sulfide bonds. The energy used to bind two sulfur molecules is thus freed and available to vent-dwelling animals. But here the resemblance ends. In photosynthesis the sun's energy harnesses carbon dioxide to make food. In sulfur bacteria the carbon dioxide is used, albeit secondarily, to produce energy in the absence of sunlight to make food.

The similarity between the two aquatic worlds—one darker than night, the other as bright as transparent water and unfettered sunlight can be—is defined by the interdependence between mutualistic participants. There is no ecological resemblance beyond the fact that both communities have independently evolved relationships that depend on invaders from the outside that live inside their bodies.

From the sea's surface to its abyssal bottom, mutualistic relationships are fundamental to spectacular, vast, and diverse communities.

THE OLDEST ANIMAL

Imagine an alien, utterly black world populated by animals that cannot breathe, for there is no oxygen. That cannot eat, for they have no mouths. That cannot digest food, for they have no digestive apparatus. It is so cold that normal enzymes cannot function. But this inhospitable environment is not light years away. This is *inner* space, the bottom of the abyss, the most nether depths of the sea. The extraordinary wormlike animal is blood-red, signifying the presence of hemoglobin, the carrier of oxygen, but there is no oxygen. It has many tentacles, yet there is no prey to grasp. It lives in a long, ivory-like tube, but there are few enemies that require it to protect itself. This animal is *Riftia pachyptila*, unknown to humans until the mid-twentieth century.

It is a worm, yet much of its body lacks the segments of the annelids; it lacks the flexible cuticle of the roundworms; it lacks the double-layered flat anatomy of the flatworms. It is so extraordinary that a new class of phylum annelida was created for it and its few relatives—pogonophora.*

Pogonophorans amaze us because they outdo the coral reef in their reliance on huge populations of internal mutualistic symbionts for their very existence. There are corals that survive without zooxanthellae—proving that they can utilize external resources like zooplankton as their exclusive food source. But the pogonophorans rely on their bacterial symbionts for the very energy of life. Inside the body cavity is an elongate organ called a trophosome, teeming with sulfur-eating bacteria. The plume of tentacles absorbs hydrogen sulfide and methane, and red blood carries these molecules to the trophosome. The bacteria use these materials to feed themselves and reproduce rampantly. What does *Riftia* get out of it? It eats the bacteria. But where does it get the bacteria? *Riftia*'s microscopic larvae are born bacteria-free. They swim and hunt like ordinary aquatic larvae and have the necessary gut, equipped with mouth and anus. They seem to be able to travel though the inky blackness from hot spot to hot spot, there to settle. But, although they have survived by attacking and eating prey, it has recently been shown that the precious sulfur-metabolizing mutualistic bacteria are not ingested; the animals are infected through their skin.

Pogonophorans exhibit several firsts: they may be the fastest growers on Earth. One species grew thirty-four inches in one year. Not bad for an animal little more than an inch wide.

Then again, they may be the oldest animals on Earth. The animals depicted in plate 25 may be 25,000 to 250,000 years old, depending on who is counting.[32] This is conjectural, of course, but it is based

---

*The phylum was described and named in 1900 from a single worm torn from the bottom two miles under the surface off Indonesia. One man studied this specimen for over fifty years, but it was incomplete; the rear portion was torn off in the collection process. In 1977 a whole worm was recovered. The torn-off part was segmented like an earthworm. It has the bristles of a sea worm. It has a fully formed circulatory system and separate sexes. Its tube is made of that cleverly invented invertebrate plastic called chitin, characteristic of insects. In fact, the class seemed to be a cross between worms (annelids) and their evolutionary successors, arthropods. After much taxonomic confusion, it is now considered a family of annelid worms, siboglinidae, contained in the pogonophora (now called subclass vestimentifera).[31]

PLATE 25

UPPER

CORAL REEF BATHED IN INTENSE SUNLIGHT. *Right:* Pillar coral, *Dendrogyra cylindrus,* with crinoid on top; *middle:* staghorn, elkhorn coral, *Acropora cervicornis, A. palmata; left:* sea fans, *Gorgonia flabellum,* with school of fishes. Corals use internal mutualistic protozoans, zooxanthellae, to convert carbon dioxide to larger carbon molecules (food) by photosynthesis.

LOWER

THERMAL VENT IN PERPETUAL DARKNESS. *Right:* Pogonophoran (vestimentiferan) tubeworm, *Riftia pachyptila,* to 6 feet by 1 inch wide; white annulated tubes contain blood-red worm. No mouth; tentacular plume on top absorbs sulfur compounds, carbon dioxide, and methane from water; hemoglobin in blood carries these compounds to mutualistic bacteria infecting tissues of major organ, trophosome. Here bacteria convert carbon dioxide to larger carbon molecules (food) by chemosynthesis. Bacteria are eaten directly or secrete food into *Riftia*'s tissues. Posterior of worm bears segments with bristles like annelids.

Above *Riftia* is deep sea grenadier fish, *Coryphaenoides acrolepsis.* Below *Riftia* are footwide clams, *Calyptogena magnifica,* and huge white crabs, *Bythograea thermydron.*

To left of *Riftia* is a "black smoker" spewing minerals, carbon dioxide, methane, and sulfur compounds carried by intensely hot water from bowels of earth. Behind black smoker is thermal vent, a crack in the ocean's bottom, also releasing these compounds at intense heat. Unseen behind *Riftia,* at far left, are footwide gray ribbed mussels, *Bathymodiolus thermophilus.* All vent animals use sulfur-based energy pathways.

on this logic: the base of a tube, arising from a hard surface, may be buried in subsequent sediment to a depth of six or more inches. Yet the deepest bottom sediments are remnants of the slow rain of tests (shells) of shallow-water microscopic animals. The sedimentation rate can be a quarter-inch per thousand years, or less. Do the math.

26

A DAY IN THE CARIBBEAN

THE ENORMOUS TURTLE PADDLES by in what seems to be an endless circle. Its lidless eye glares at me, but although a few feet away, its interest is focused on the blurry distance. It seems oblivious as I touch the three-foot-wide shiny shell. Three nickel-size camouflaged conical projections jut almost imperceptibly from the carapace, scarcely visible except to my expectant eye. These are turtle barnacles, destined to spend their lives carried from place to place on the back of their ever-moving universe. "Why are they adapted only to life on a turtle's back?" I mused. Then I realized that it was not the bits and pieces of the turtle's lunch that the barnacles were after. They were in it purely for the bus ride. Movement is the name of the game. Stretching jointed, hairy legs into the almost empty sea in their quest for tiny motes of life to strain from the water is hard work in this Caribbean aquatic desert. There are places where plankton is abundant, such as in mile-wide gyres, whirlpools in the ocean that sweep random microscopic swimmers into reach. The turtles swim from place to place, eventually bathing the barnacles in the precious life-sustaining cloud of soupy food.

The commensal relationship where a partner is in it only for the ride has a special name, phoresis.

The turtle ponderously moves toward a shadow. A fisherman has returned to his shucking spot in his one-masted, gray-sailed boat and is throwing conch guts over the side. I surface to see him whack the claw end of a hammer at a crucial spot on the pinkly glowing shell. The giant snail's body slips slimily out of the shell to the boat's bottom. He picks up the mess and cuts the guts from the snail's foot, severing flesh from internal organs. The flesh is destined to be the makings of conch fritters and soup served in the hotels visible as colorful cubes on shore. The guts go over the side, becoming manna raining down on hordes of fishes—that is, if the turtle doesn't grab them, gulping the mess with its beak and leaving a brown cloud of liver bits that attract a halolike horde of tiny, shiny, blue and yellow fishes.

The turtle circles expectantly, waiting for the next gift from heaven. As I watch, the turtle angles upward. A soft, brown protuberance almost a foot long extends from the yellow bottom of its carapace. Was this an extraordinarily well-endowed animal? Was this his giant penis? I didn't think so. Suddenly the brown projection begins to undulate. It separates from the turtle! It is a sharksucker, *Echineis naucrates*, and it is doing its thing, gobbling up scattered remnants of the host's meal. Sharksuckers don't stick only to sharks. I have seen them attached to fishes scarcely larger than themselves. In fact, later that day I watched a kid stare as a fishy projectile shot up from nowhere and nuzzled up to his leg. He fearfully shook it off, horrified at this close encounter with a sharksucker.

As the turtle returns to feed, I look expectantly for a turtle leech projecting from its neck. No luck.

Turtles are a veritable cornucopia of external symbionts, more attractive than most other aquatic animals because they move slowly and have a lifelong rigid shell. But with the exception of the parasitic leech, they use it for transport only. The turtle is not a source of sustenance, just transport, a moving rock. Are the symbionts adapted to their special habitat? Before you decide, consider that the barnacles often have unique T-shaped flanges that facilitate attachment. There are sixteen species found nowhere else but on turtles. The sharksucker has a large, corrugated dorsal sucker.

A ROMP AROUND THE REEF

This was one of those exquisite Caribbean days, limpid seas reflecting the looming Jamaican mountains. A palette of bright green splashed with darkly flowing shadows. White sunlit beaches were edged with dark, serrated palm trees, their leaves floating above distant, slender trunks.

The tepid water was as clear as air, and I couldn't resist diving to the bottom and swimming with my belly almost touching sharp corals nestled among green leaves of turtle grass.

I reached the reef, whose vast, voluptuous variability invited penetration. Shimmering golden and greenish mounds and sharp orange-green stag and elkhorn corals enveloped me. Black holes, caves, and crannies, alluringly mysterious, cried out to be investigated. By arching my body and kicking hard, I thrust myself down eight feet, where, by holding on to dead coral edges, I could peer into the blackness until, finned feet unwillingly floating upward, I was pulled away. Again and again I fought the good fight, grasping chunks of greenish fuzzy dead coral until, after a minute or two, I was out of breath.* The search continued. A school of blue striped-grunts, *Haemulon sciurus*, crowd together in fear, forming a thick yellow-blue cloud. A huge red, blue, and green blunt-browed, bovine supermale stoplight parrotfish, *Sparisoma viride*, grazed on corals with beaklike buck teeth. The scraping sound from each bite was amazingly audible as he rasped off the polyp-covered, calcareous topmost layer of a coral mound. A cloud of chalky, ground-up coral spurted from his anus.

The bottom was streaked with bright evanescent lines, reflections of tiny rippling waves. All the conditions were right. This was going to be a good snorkeling day.

MORNING

Randomly wandering from spectacular scene to scene, I came upon a vague, staggered lineup of six large fishes, each hanging in the water in a peculiar, unnatural pose—one with head pointing upward, yawning; another, head facing downward with fins extended. One hovered with its body dangling diagonally. I furtively moved to the end of the line. This was a "barber shop," and I tried to be just another

---

*Nonbiologists should never touch living corals. You may not know what to grasp and thus harm the reef.

customer. Just close enough to be visible, a coral head was the focus of the line. On its pinnacle, disporting itself for all to see, was a two-inch-long barber pole shrimp, *Stenopus hispidus*. It danced its dance of life—and death, for if it stopped its inviting undulations it would immediately be snapped up by the predatory fish it was entertaining. Instead, it leaped onto the flanks of the first fishy customer, rapidly darting toward the gills, stopping briefly along the way to pick off a parasitic copepod or two. It entered the gill chamber. After a minute or so it emerged from the mouth, picking the fierce teeth clean. Finished, it slowly floated downward to the pinnacle of coral, its "cleaning station."[33]

Again it danced, twirling its red-and-white striped body to invite the next customer. I longed to be cleaned along with the rest of the customers, but I knew that was not to be. I couldn't produce the appropriate signal to initiate the process. The snorkel in my mouth precluded the distinctive "yawn" that would invite the cleaner to clean my teeth. I can tell you how it feels though, because we cultured cleaning gobies in my aquaculture lab. When I immersed my finger into a tank of hundreds of them they converged in a jostling, quivering mass. They didn't need the traditional stimulus in this artificial milieu. Less than gentle, their tiny pecks left my cuticle and nail immaculate. I was a partner in a mutualistic relationship!*

In the sea, this inborn, highly evolved interaction between the animals consists of a genetically imprinted dance performed by the cleaner and recognized by the customer, who turns off its predatory instinct and assumes its genetically programmed acceptance response, hovering in an incongruous position. Upon seeing this, a cleaner shrimp waves long white antennae and mounts the fish to remove its parasites, benefiting its host and gaining a good meal, for a fish can have hundreds of parasites on its flanks or sucking its blood from thinly protected gills.

---

*This word, mutualism, has long superseded the word "symbiosis" to label this mutually beneficial relationship. Symbiosis means "living together." All-encompassing, it refers to all three of the primary interactive trophic behaviors: mutualism, parasitism, and commensalism.

This complex interaction of dance and posture is a brainless phenomenon. Fish and shrimp inherit a specific complementary performance, a dance of the genes, programmed in the distant past. How can such a complex relationship of behaviors evolve? It almost makes one believe in "intelligent design." But such an explanation requires accepting an idea on faith. It is not satisfying to the logical mind. An empirical explanation is necessary. Viewed from the biologist's perspective, and using parallel evidence from other behaviors like nest building in birds or the intricacy of a spider's web, we can make a rational explanation.* Here is one possibility:

A tiny goby or shrimp is grazing on a mound of coral. Finished with its rocklike pasture, it moves on to another copepod-covered looming mound—only this is the broad flank of a large fish. One gobble and the shrimp is no more. Over eons some otherwise predatory fishes didn't suck in their tiny prey. Perhaps their skin was not sensitive; perhaps they couldn't or wouldn't eat the little pest. The fortunate shrimp fed on its host's parasites. This cleaned fish survived. Its congeners became exhausted by the horde of parasites occluding their oxygen-providing gills and tearing at their skin. (On the Great Barrier Reef, field studies demonstrate that fishes become ragged and weakened if they are not cleaned. Fewer large fishes are found in cleaner-deficient segments of the reef than on areas with normal populations of cleaners.)

The cleaned fish lived; many of its neighbors didn't. Soon this genetic capacity to retain a cleaner proved to be so beneficial that the mutated host fish replaced the others of its species. More eons elapsed; the predator evolved another behavior. It somehow encouraged the cleaner to mount it. Over millions of years a set of behaviors was woven together, an intimate dance, predator dancing with prey. When the dance ends, the truce is over. After a short time interval, its trance disappears and the predator fish is ready to engulf its usual shrimpy prey. But the barber receives its "tip." The predator has floated far from the cleaning station before awakening. The shrimp is safe.

---

*If a spider is hatched in a box and never sees another spider, it will still spin the characteristic web of its species. The odds of a rarely occurring phenomenon happening once become better and better with each repetition until it becomes inevitable.

AFTERNOON

The glaring sun was now at its zenith. Any sensible animal would be hiding in the shade of coral overhangs. I searched in dark holes. A flash of red. A black-barred soldier fish, *Myripristis jacobus*, glared at me with huge lidless eyes. Something was strange about these eyes. As it turned to disappear into the dark inner recesses under the coral head, I could see a huge, flat, buglike crustacean, the isopod *Anilocra laticaudata*, above one eye.

It all starts when an isopod larva lands somewhere on the fish. During a short period of development it migrates forward, always ending up near an eye, where it takes up residence like a huge, bulbous, segmented tumor. Always near the eye.

All *Anilocra* juveniles start off as tiny males and then become inch-long females that dwarf the eye of their host. They are large enough to brood a hundred eggs in a ventral chamber called a marsupium, like a kangaroo. One large gravid female per fish is the rule. Why? In a not-unusual mechanism of population control, the presence of the female inhibits the development of the numerous males on the fish. None will become a female unless she dies. This eliminates competition and assures there are tiny male sexual partners nearby. If she dies and her chemical influence declines, an immature male soon becomes a female and replaces her. So his sexual identity is not controlled internally but is governed by the emanations of the female.

Your childhood "ball bugs" or "pill bugs" are terrestrial isopods (iso = same, pod = leg). Their seven pairs of legs are identical. They are harmless. But these aquatic parasitic isopods have piercing mouth-parts to suck up blood and tissue and sharp claws that cling so fiercely to the host fish that they create necrotic lesions, inviting infection, endangering it. Killing the goose that lays the golden egg is against host-parasite rules. This suggests that this fatal relationship is relatively recent.

Is the isopod-fish relationship different from the barnacle-turtle relationship mentioned above, or part of a continuum?

PREDATOR OR PARASITE?

Absorbed with thoughts of isopod evolution, I drifted downwind into a veritable garden of soft corals. Bright purple and yellow sea fans, *Gorgonia flabellum,* and six-foot-high sea feathers, *Pseudopterogorgia americana,* enveloped me. I was floating, weightless, at peace. The current carried me. Flexible "feathers" caressed my naked skin. A splash of orange on a bright purple pallet attracted my eye. I plucked the colorful bump off the sea fan. The orange-spotted surface shrank inside a longitudinal groove and I was left holding a pale purplish shell. This is the most exquisite of all Caribbean snails, the flamingo tongue, *Cyphoma gibbosum.* Startled, it withdrew its leopard-spotted fleshy mantle to become an innocuous pale shell, a defense mechanism that saves it from tourist collectors. Who wants a dull shell?

The flamingo tongue lives preferentially on sea fans and occasionally on related soft corals, making it somewhat host specific. It grazes on the living polyps that invisibly cover the fan's surface. Is it a predator or a parasite? When alone, it fulfills the definition of a parasite, for the trail of death it creates in its wake is quickly closed by regrowing polyps and the duo lives a balanced existence. But when a pair or trio of flamingo tongues descends on a sea fan, their depredations overwhelm the fan and it dies. A predator is usually larger than its prey; parasites, smaller. Flamingo tongues are smaller. A predator quickly kills its prey. In slow motion, several flamingo tongues kill theirs.

Predator or parasite?

EVENING

Snorkeling home, back baked red and exhausted, I stopped abruptly. One last underwater tête à tête. Almost hidden in the turtle grass was a C-shaped, brown, loaflike, foot-long five-toothed sea cucumber, *Actinopyga agassizii,* supine, sweeping sand into its mouth with twenty-five small, brushlike tentacles. Its anus was bedecked with five square teeth. No teeth in sand-filled mouth, but teeth in anus. Weird. I picked it up, feeling its heaviness and rough exterior, turning it over to see

PLATE 26

A. GREEN SEA TURTLE, *Chelonia mydas.* To 5 feet, largest of sea turtles. Teardrop- shaped carapace with white line on edge; shiny brown with yellowish striations on each section. Green flesh. Adults have single claw on legs; mostly vegetarian. Kept alive on ships for months because they don't move but stay alive when stored on their backs. Endangered; eggs eaten. Can dive for four to five minutes, breathe for one to three seconds.

B. SHARKSUCKER, *Echineis naucrates.* To 29 inches; light gray with black stripe that starts near eye, broadens on abdomen, tapers toward tail; grooved plate on head; upper jaw shorter than lower; dorsal and ventral fins mirror images. Feeds on turtle's scraps.

C. BARBER POLE SHRIMP *Stenopus hispidus.* To 3 inches, white banded with red; claws on first three legs; third pair longest; very long antennae; has set up cleaning station on coral head, dancing enticingly. Fishy customers at odd angles signifying lack of aggressive intent.

D. FLAMINGO TONGUE, *Cyphoma gibbosum.* To 1 inch; snail has black-edged orange spots on mantle that can be withdrawn, revealing purplish shiny shell with vertical aperture. Eats polyps of purple sea fan, *Gorgonia flabellum.*

E. PEARLFISH, *Carapus bermudensis.* To 7 inches; pearly pink, no fins; tapers to posterior point; in anus of five-toothed sea cucumber, *Actinopyga agassizii;* to 1 foot; brown on top, white under; five rows of tube feet; five teeth in anus, no teeth in mouth.

F. PARASITIC ISOPOD *Anilocra laticaudata.* To 1 inch; gray segmented body; seven pairs of legs; damages skin on head of various fishes; fish can die of bacterial infection. Blackbar soldierfish, *Myripristis jacobus,* shown, to 10 inches, red; vertical black bar along gill, large eyes, nocturnal.

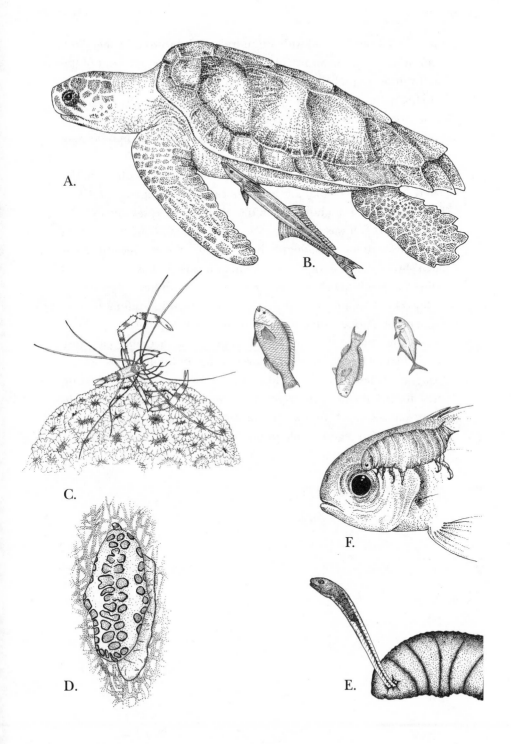

A.

B.

C.

D.

F.

E.

the pale bottom, pocked with a hundred small brown tube feet. Suddenly a sleek, iridescent fish about six inches long popped out of the toothy anus. I grabbed at it but it slipped from my fingers, dropped into the water, and was gone in a silvery flash. This was the pearl fish, *Carapus bermudensis*. It lives in the rectal region (cloacal chamber) and body cavity of this particular sea cucumber (rarely found in other species). The fish leaves its rectal haven at night and returns in the morning like a vampire returning to its coffin. But any anal coffin will do. There is no evidence that the adult fish can find its original *Actinopyga* host, but its larva can home in on the smell of *Actinopyga* mucus.

The pearl fish was thought of as a commensal, obtaining benefit from the cucumber host but not harming the cuke. But recently it has been shown that it sometimes rips out and eats chunks of the cloaca wall. Change of status! Once commensal, now parasite.

Still clutching the cucumber, I looked again at its bottom. In a fitting end of a wonderful day, I saw six tiny, sharply pointed, half-inch-long white snails, *Odostomia*, highly adapted sea cucumber parasites. No ambiguity here! The snails have lost the traditional snail rasping "tongue." It has become modified into a tube used like a drinking straw to suck the body fluids from its host. The snail has made an irreversible commitment and cannot survive like an ordinary snail. It must be able to digest a hole in the sea cucumber's thick armor and penetrate into its body with its long, tubelike proboscis to survive. No other host will do.

The end of a perfect day.

27

TIT, TIT, TITTIE—CUCKOO

"TODAY'S LECTURE WILL COVER THE SUBJECT 'TITS.'"

Judging by the smirks and sidelong glances of the hormone-infused students, the class is interested. How to keep them involved?

"And cuckoos." A few of the brightest realize I am talking about birds and promptly go to sleep. The others maintain their smirks a little longer.

CUCKOO—OR CLEVER?

The British have a penchant for coining peculiar words. Take, for example, "bangers and mash" (sausages and potatoes) or "loo" (bathroom). Oddly, they call small birds of the genus *Parus* tits. A larger bird, *Clamator glandarius*, with a mellifluous call is a cuckoo.* The two species are intricately involved in a symbiotic relationship called brood parasitism.

---

*The female cuckoo is famous for leaving its eggs in the nests of other birds. If we didn't know better, we would think that the cuckoo's habit of leaving its eggs in another bird's nest is stupid. But early biologists didn't know it is diabolically clever. Further, the sound of the bird's call seems, to some, like inarticulate speech, odd or foolish. The cuckoo's name in early German was *gauk*. This word turned up in early German texts meaning "fool" or, later, "idiot," hence "being *cuckoo*" means behaving foolishly. The word "to gawk," meaning to stare foolishly at something, has a similar origin.

Americans, too, create bizarre names for birds. The brown-headed cowbird, *Molothrus ater*, has no obligate permanent relationship with cows, nor does it resemble one, nor does it have tits.*

The cuckoo, the cowbird, and smaller birds like tits interact in a behavioral manner as intimate and host-specific as any physiological parasite-host interaction.

It has long been known that cuckoos and tits have a one-way relationship. The cuckoo tricks the tit into rearing its young. The smaller bird, having sparse room for its bird-brain in its tiny head, seems particularly dumb. The cuckoo's bird-brain is particularly unscrupulous.

Upon discovering an active tit nest, the cuckoo promptly lays a huge egg among the three or four tiny eggs in the nest and flies off. The parent tits never notice the difference. They brood all the eggs. But in this evil, eons-long interaction, the cuckoo egg hatches first, producing a ravenous, open-mouthed infant, constantly squalling for food. The parent tits feed the miniature monster. Then, belatedly, the tit eggs hatch helpless, hapless young. The alien infant thrashes around violently, throwing out one by one the competing, poorly developed juvenile tits. The parents exhaust themselves bringing worms and insects to thrust into the always-open maw of the juvenile cuckoo. After an abbreviated fledging period, the successful parasite flies off to continue the cycle. In this instance of sibling rivalry, the elder sib has wreaked such havoc on its nest-mates as to make human siblings seem like angels.

This devilish demonstration of birdy immorality has a recently discovered nasty variation. Not content with molesting the tit with its avian scam, the cuckoo tries its nefarious scheme on the magpie, *Pica* spp. Magpies are known for their intelligence. Not dummies like tits (some have even been taught to "speak"), these birds are not going to

---

* Cowbirds graze on insects in open fields like cattle egrets, *Bubulcus ibis*, using the cows to disturb insects on which they feed. But often they are found in small flocks on lawns or fields, without cattle. The association with cows is ephemeral and vague. African red-billed oxpeckers, *Buphagus erythrorhynchus*, on the other hand, are often considered to be mutualistic, obtaining nourishment from ticks and botfly larvae on cattle. However, they also peck at wounds on the cow's back, keeping them open, attracting more flies. Thus they can ultimately be considered parasites.

have something put over on them. They readily throw the cuckoo egg out of the nest. Faced with this formidable intelligence, the not-so-cuckoo spotted cuckoo has had to resort to brute force. In response to the clever rejection of its eggs, it has evolved a mafia-style protection racket. A Tony Soprano-style "enforcer" is brought in. The birdy thug returns and destroys the magpie eggs or kills its chicks.

If it knows what's good for it, when the magpie lays its next brood soon after, it will accept its parasitic burden.

EMPTY NEST SYNDROME

Cowbirds, *Molothrus ater*, are larger than robins, and prothonotary warblers, *Prothonotaria citrea*, are smaller. When a particularly observant warbler notices that it can hardly get its small body around the cowbird egg among its own little eggs and rejects it, the cowbird returns and maliciously breaks the warbler eggs.* This, as we have seen, is the bullying mode of a brood parasite.

Warblers lay their eggs early in the spring before the cowbirds are ready to lay theirs, depositing eggs randomly over a month or more. This enhances survival prospects for the species. A long reproductive time spectrum allows an organism to extend its period of fecundity to survive short-term environmental vicissitudes. (If a species of plant flowers all summer, its seeds will survive periods of drought and predation.)

When their fledglings have been displaced by a cowbird, the warblers respond to hormonal mandates and exhibit an "empty nest syndrome." They will automatically lay another batch of eggs. The cowbirds monitor the new nest and *at the appropriate time* return to lay their monster egg in the poor warbler's nest. Sad.

How does this morality play end? Will the evil cowbird win out? New data reveal that on the average, 20 percent of warbler nests are

---

*While European brood parasitic birds produce eggs similar in color to those of their hosts, cowbirds produce four or five white speckled eggs that are different in color and size from the more than two hundred species of songbirds they parasitize. It has been hypothesized that cowbirds have evolved brood parasitism because they followed migrating bison herds and had no time to build nests.

PLATE 27

BROOD PARASITISM

A. GREAT SPOTTED CUCKOO, *Clamator glandarius*. Male to 13½ inches; dark gray, white belly; gray pointed cap; gray wings; yellowish face. Lays up to thirteen tan, ochre-flecked eggs in nest of smaller coal tit among tit's seven to eleven red-spotted, white eggs. Cuckoo eggs hatch four or five days sooner than tit's. Obligate parasite; does not build its own nest.

COAL TIT, *Parus ater*. Male to 4½ inches, black head with large white patch on top, neck, face; wing gray; belly whitish; bill black.

B. NEST OF COAL TIT fledgling cuckoo, hatched earlier, is larger by virtue of larger size of species and the fact that it has been fed for four or five days; thrashes around and throws tit's young out of nest or pecks them to death.

C. BROWN-HEADED COWBIRD, *Molothrus ater*. Male to 9 inches; brown head, metallic green-black body. Usually lays one gray spotted egg among four to six lighter-colored host eggs. Female can lay thirty-six eggs in one season. Only brood parasite in North America. Lays eggs in nests of over two hundred species of birds. Some birds reject eggs by (1) abandoning nest; (2) destroying egg; (3) throwing fledgling out of nest.

PROTHONATORY WARBLER, *Prothonotaria citria*. Male to 4¾ inches; yellow with gray wings; wide gray tail has large white patches; long black bill. Female lays four to five eggs; young remain in nest after hatching twelve to fourteen days; vulnerable because of smaller size. Named after officials in Catholic Church, prothonotarii, who wore golden robes.

A.

B.

C.

parasitized by cowbirds. One would expect the same 20 percent of second nests to be attacked, but new investigations reveal that an amazing 85 percent of the second nests were attacked. Clearly, the second nests were being monitored by the cowbirds. These data are convincing. But so are these: the warbler parents produce, on average, three surviving juveniles when parasitized, but only one when left alone. The warblers actually fare better from their encounter with their brood parasites!

28

THE GAME OF LIFE:
NAME THAT CATEGORY

CAN YOU DECIDE which of the following symbiotic relationships is parasitism, commensalism, or mutualism?

THE PERFECT WEDDING GIFT

Consider the relationship between the hollow, tubelike Venus' flower basket glass sponge, *Euplectella*, and its aptly named shrimp inhabitant, *Spongicola*. When its cellular component is removed, the sponge appears to be an exquisitely woven cylinder of glass fibers. Through the network of large and small openings, a pair of shrimps can be seen. They entered as a tiny couple able to squeeze through the sponge's larger pores. Feeding on plankton swept into the central chamber of the sponge, the couple grows and prospers until too large to escape. The sponge provides all the necessities, and the happy couple release their fertilized eggs through its innumerable pores.

Ah, love. The invertebrate component of romance is instinctive and emotionless. But the paired shrimp have inspired overly emotional Japanese to indulge in a romantic ritual. Upon their marriage, the glowing human couple is given a glass sponge with its shrimp inside to symbolize the eternal union of the bride and groom. (Unfortunately, some cynics have interpreted the locked-in aspect as a symbol of the eternal entrapment of marriage.)

Parasitism, commensalism, or mutualism?

A MOVING RELATIONSHIP

A more "moving" relationship involving sponges is the romance between crabs of the genera *Dromia* and *Dromidia* and a variety of sponge species. The crab removes snippets of sponge with its long, pincerlike claws and places them on its back. The sponge regenerates and grows over the crab. Predators do not notice the slow movements of what appears to be a brown, formless sponge. Few predators eat sponges. Mixed in with their network of calcareous spicules and tasteless fibers are a series of noxious compounds—but everyone likes to eat crabs.

This is clearly a long-term evolutionary interaction because the crab has physical adaptations. Its back is covered with hairlike curved spines that serve as a perfect substratum for the sponge to regenerate into a replica of its original amorphous self. Eventually the crab is overgrown by the sponge. No crabby features are visible under this walking sponge mound. The crab is safe.

Parasitism, commensalism, or mutualism?

THE RHINOCEROS AND THE ANTELOPE: A PARABLE

Once, on safari, the guide took us to see an unusual phenomenon, an interaction between the largest African antelope, an eland, *Taurotragus oryx,* and a black rhinoceros, *Diceros bicornis.* The eland stood about twenty feet from the grazing rhinoceros. This unique combination had remained stable for several months, perhaps years, before the guide stumbled on it. The guide conjectured that both animals received benefits: the rhinoceros, vulnerable because of its weak vision, would be warned by the sharp-eyed eland at the approach of predators before they could endanger both animals. The formidable rhino would charge at the blur indicated by the eland. Is this suggested symbiosis real or imaginary?

Parasitism, commensalism, or mutualism?

WHEN A FISH BITES YOUR HAND, THAT'S A MORAY*

Standing in the stern of the boat, the instructor intoned, "This is a spectacular site; you will see an extraordinary number and variety of fishes here. Why? Because a conch fisherman comes to this exact spot every day, yanks out most of the meat in his conchs, and throws the shells of these huge edible snails overboard." The bottom was covered with giant shells, their freshness attested to by their unsullied, glossy pink apertures. They were lying on a pyramid of algae-encrusted remnants of former harvests. The professor snorkeled to the bottom and found a clear space. Gesticulating, he began to point out crevices and holes and their hidden inhabitants. Absorbed with his grunting underwater lecture, he rested one hand on the soft sand near some shells and gesticulated toward the class with the other. Suddenly he felt a sharp, piercing pain. A flood of blood poured from his pinky. The attacker was visible in the shadows. It was among the most fearsome of fishes, a moray eel, *Gymnothorax*. Razor-sharp teeth had severed muscle and tendon so extensively that the wound didn't heal right. To this day the instructor's pinky is frozen in a grotesque L.

Moray eels lead solitary lives hidden in crevices. They are scary, but dangerous only to divers who put their hands into dark holes—and this instructor who confronted an unusually irritable moray.

A newly suggested relationship has been described between the shy, normally hidden giant moray eel, *Gymnothorax javanicus*, and a species of grouper, *Plectropomus pessuliferus*, a huge, predatory fish that cruises the periphery of the reef.

---

*Parody of a song title of the 1950s, "That's Amore," sung by a fellow Brooklynite, Dino Martinelli (Dean Martin).

PLATE 28

UNDERWATER SCENE

PURPORTED RELATIONSHIP BETWEEN GROUPER AND MORAY. Grouper (upper fish) signals moray to pair up; both scout reef; when a prey fish appears, grouper scares it into crevice; moray follows and either scares it out (whereupon it is promptly eaten by grouper) or eats it.

A. CORALGROUPER, *Plectropomus pessuliferus*. Huge, to 47 inches; gray with ocher bars, iridescent blue dots; large teeth.

B. GIANT JAVA MORAY, *Gymnothorax javanicus*. Gigantic, to 9 feet, thick as a man's thigh; brownish with dark brown spots, black blotch at round gill opening.

C. VENUS' FLOWER BASKET, *Euplectella aspergillum*. To 16 inches; sponge; in nature brown tapering tube (shown); when tissues removed, becomes lacy cornucopia made of beautiful glassy threads; sievelike plate closes top. Many large openings interspersed among tiny pin-holes. A pair of shrimp, *Spongicola japonica*, enters when juveniles, mate; larvae escape through holes. Adults grow too large to escape, feed on particles carried inside by flagellated circulation of sponge.

D. SPONGE CRAB, *Dromidia* spp. To 3 inches; brown with red claws. Spiny fuzz on back; last pair of legs dorsal, used to hold en-crusting organisms like sponge on back until they can overgrow body. Here, brown sponge camouflages body.

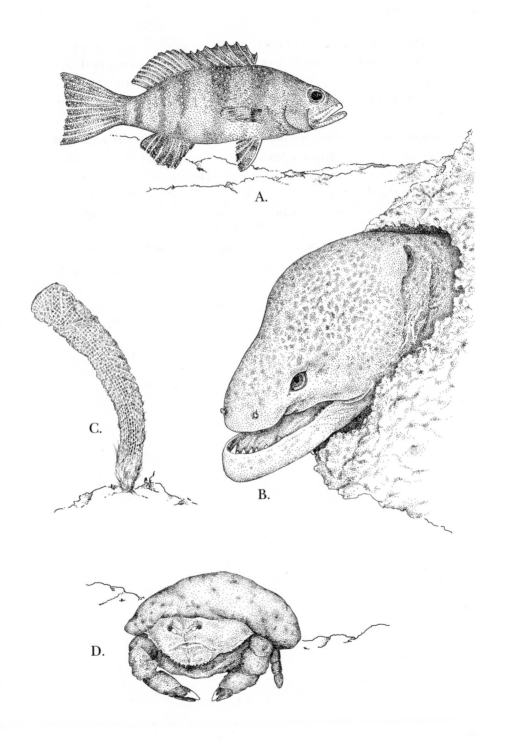

A.

B.

C.

D.

The grouper sees a moray's eyes glaring from a coral crevice. It shakes its head violently. Immediately the moray swims out of its lair and follows the grouper to a crevice too small for the grouper to enter. The moray slithers into the hole and eats the fish that the grouper has scared into hiding. Occasionally the small fish escapes the moray and the grouper gulps it up. The authors of the study suggest that this relationship is possible because the two participants are so large that they swallow their prey whole, leaving no carcass over which they might fight. The researchers checked the stomach contents of many pairs in this putative interaction, and each contained more prey than normal. (Grouper stomachs contained five times as many fish as the control group.) As further evidence, the association between the normally hidden moray and the grouper lasted for as long as forty-four minutes. Real or imaginary?

Parasitism, commensalism, or mutualism?

29

PAEAN OF PRAISE

I REACH INTO THE CARTON, newly arrived from the biological supply house, trying to keep the flaps closed with my elbow. The contents feel cold and slimy. I grasp blindly. My unseen prey squirms in my hand. I withdraw it. Uncomfortable, I move too quickly and my elbow no longer presses on the flap. It swings open from staccato blows from underneath. Out pops a horde of leopard-spotted frogs. I grab at them futilely, one hand holding a slimy frog intent on escaping my grip. It escapes. Frogs are leaping everywhere, bouncing off walls, cabinets, prep tables. I yell for my assistant. The prep room is a blur of frenzied activity. She runs one way, I run the other, clutching frogs. The class is waiting beyond the closed door.

Each frog we catch is popped into a gallon jar with a wad of chloroform-soaked cotton at the bottom. The lid is closed and the frog peacefully drifts off to perpetual sleep. As the jar is opened for each of the twelve frogs, the sickly sweet smell of chloroform seeps into the air of the closed prep room. Soon, I too am nodding off. The frogs are transferred to a tray. I stagger into lab balancing the tray and distribute the frogs into waiting dissecting pans.

I often envision armies of dead frogs marching on my soul in the afterworld, joined by legions of dead mice, seagulls, and cockroaches. Believe me, there is nothing so off-putting as this phantom horde.

PARASITE HOTEL

Frogs are a veritable treasure trove of parasites. They fully live up to my demanding standards. In addition to massive infections by the lung flukes that make frogs so dubiously famous, they are hosts of one-celled animals that can be found only in frogs and toads.

Many protozoan symbionts are found in the rectum. The blood contains the larvae of a stringy nematode similar to those causing the testicle-stretching human filarial disease, elephantiasis.

Slicing open the intestine of a frog releases a murky explosion of gut contents, a seething jungle of twisting, swimming, staggering, and crawling inhabitants. But this generous parasite-provider has more largesse to offer the curious—cysts under the skin and embedded worms hidden in the lungs.

Frogs live in a liquid medium and are vulnerable to all the symbionts that have not bridged the gap to the land, and some that have.

Two dazzling parasitic stars are available for admiration by the throngs of their fans (all three or four of us). One is the only monogenetic trematode that lives "inside" the host's body. The monogenea are a class of external parasitic worms, discussed earlier, that are distinguished by the fact that they crawl on the surface of fishes and other aquatic animals (one has even been reported on the eye of a hippopotamus). But our hero, *Polystoma integerrimum*, has chosen a peculiar home. It lives only in the urinary bladder of frogs.

Is this species a maverick, invading the inside of its host, becoming the only monogenetic trematode species that is an *endo*parasite (internal) in a class of (external) *ecto*parasites? Is it an evolutionary precursor, a leader in the transformation from outside to inside?

Some say that the urinary bladder is not technically inside the body, but I say living in the bladder is as "inside" as the lumen of the intestine, and no one doubts that a parasite living inside the gut is an endoparasite. Besides, it is enough cause for celebrity that the worm lives exclusively in urine, sucking blood from the lining of the bladder like that grisly parasitic fish, the candiru.[34]

The other unique animal is the majestic *Opalina*. Who can forget opening the rectum of a frog to see huge, ciliated protozoans regally gliding along in the rectal fluid like motile, translucent leaves?

These animals were formerly lumped in with ciliates like *Paramecium* because they too use a coating of cilia to row with, sailing along like oar-propelled ancient warships. But now it is thought that *Opalina* has so few things in common with the ciliates that a new subphylum has been erected for them. Let us not be blasé—the establishment of a new subphylum is monumental. So much fundamental taxonomy was accomplished in the eighteenth and nineteenth centuries that it is indeed cause for celebration that a new subphylum is erected.

As with *Polystoma*, its habitat is remarkably restricted. It lives only in the rectal region of frogs and toads. Talk about a limited world. Not any rectal region is satisfactory, just that of these amphibians. *Opalina* specializes in feces; the famous *Polystoma* in urine, an extraordinary sharing of the excretory wealth of frogs.

DANCING WITH HORMONES

To the human eye the lowly mud-dwelling frog is, at best, unattractive, but its role as a host to hordes of parasites of extraordinary diversity makes it a king of sorts. And its parasites live up to their regal surroundings by their bizarre life cycles.

Polystoma produces one to three eggs every ten to fifteen minutes and deposits 2,000 to 2,500 eggs during its short laying season. How short is the laying season? It starts and ends with the frog's spawning season. How does this happen? Amazingly.

Amazing #1: When spring comes and the frogs begin to spawn, the worms mate. *The frog's sexual hormones induce the worms to mate.*

Amazing #2: The eggs are expelled in the urine. They hatch, releasing swimming larvae, oncomiracidia, synchronously with tadpole development. *Only when* mature *tadpoles are available do the* Polystoma *eggs free the hairy little barrel-shaped oncomiracidia.*

Amazing #3: The oncomiracidia swim around using cilia. They will not attach to any old tadpole gills. They actively seek external gills that are ready to become internal. When the tadpole resorbs its external gills, the oncomiracidium finds itself inside the tadpole and begins to mature. It then migrates to the urinary bladder, on the way turning into an adult with an array of attachment hooks.

Amazing #4: If an oncomiracidium mistakenly settles on a young tadpole whose gills are not ready to resorb, it becomes sexually mature and produces an egg *all by itself.* The egg does not hatch until the gills resorb, releasing a female larva. In this extraordinary case, the larval worm becomes sexually mature while its body remains physically immature—so *Polystoma* has two ways of reproducing: the regular way and asexually—*an infant produces an egg!*

Opalina also has a sexual and an asexual stage, and development is also affected by the frog's hormones.

During nonbreeding season (autumn, winter), the host frog lives on land. In spring thoughts of love inspire the frogs to become aquatic. All winter only huge (visible to the naked eye), leaflike adult *Opalina* swim around in the frog's rectum. When the frog mates and lays eggs in the spring, the adult *Opalina* divide into many small forms that encyst.

The cysts pass out of the frog into the water and are eaten by tadpoles. These become eggs and sperms, and fertilization occurs in the tadpole's gut. *Opalina sex* occurs just after frog sex. Thus *Opalina*'s reproductive cycle is synchronous with the frog reproductive cycle and is controlled by the frog's hormones as in *Polystoma.*

Coordination with host behavior is necessary because the host frogs spend part of their lives in water and part on land. Both parasites need water to be transmitted to new hosts. Survival of the parasites depends on mimicking the reproductive cycle of their hosts. The hosts hibernate on land. Their sluggish metabolism is not conducive to *Opalina* reproduction, so the protozoans wait until the raging hormones of spring signal a high metabolic rate in the host frog. Ain't evolution grand?*

---

*The parasite must evolve along with its host. When the host changes, the parasite must adapt or become extinct.

GENDER BENDER

The weirdness continues. How about a hermaphrodite whose body contains both sets of sexual organs—but not at the same time? First, you have a run-of-the-mill macho male, complete with a set of male organs. Then, as time goes on and he grows larger, his male apparatus begins to shrink and is gradually replaced by female gonads. *The small male turns into a large female.*

Why large females? In territorial animals the male is larger than the female. He needs to defend his lek, his reproductive territory, from intruding males. He needs to defend his mate. He needs to prevent other males from inserting their genes into his female, otherwise, he would be supporting someone else's genome. In many cases, such as in tilapia (a fish), the female chooses larger, stronger males, instinctively "knowing" that selection of a large male mate assures her of strong offspring.

But parasites have no lek to defend, and females are larger. Why? Consider the facts: males produce tiny sperms. Females must carry thousands of huge eggs—in some cases tens of thousands. So as time goes on and the male grows, his masculine organs begin to be replaced by egg precursors, and finally where there were testes now there are ovaries. This sequence is called protandrous hermaphroditism (pro = before, androus = male).

The nematode lung parasites of frogs, *Rhabdias bufonis* and *R. ranarum*, are protandrous hermaphrodites. First, the small male produces sperms that are stored in a seminal receptacle. Time and growth go on. Eventually he fertilizes himself (or herself) with the saved sperms. But there is a futility to this procedure no matter how much fun it is. Since eggs and sperms come from the same organism, reproduction does not provide for the variability that is the stuff of evolution. How could this species survive over time in the changing environment inside the gut of the frog?

Amazingly, the eggs hatch in the gut to produce a generation of "regular" males and females. They pass out of the frog's body and are noninfective, living and feeding on bacteria in the rich, organic soil surrounding ponds. They mate. The small males die. The now-pregnant female contains thousands of larvae. *But she can't give birth.* She has no birth pore. What to do? Thousands of larvae are squirming in her uterus, getting bigger all the time. No birth. Instead, her massive brood bursts from her body, squirming all over the place, surviving for months by eating organic material in the soil.

As time goes on, some of the larvae stop eating and undergo subtle body changes, becoming infective. A frog passes by. It rests in the moist mud so conducive to larval nematode survival. Infective larvae burrow into the frog's vulnerable webbed foot, enter its bloodstream, and end up in its lungs.

Here in the lung the female *Rhabdias* begins the process that will end with her distant daughter's self-destructive explosion of larvae (see above). Mom lays eggs that are carried out of the lungs by reverse-beating cilia lining the tiny air tubes of the frog's bronchial network. These tubes become larger and larger, and the *Rhabdias* eggs flow upward in a ciliary sea into the throat and are swallowed. On their way to the intestines, the larvae mature and, on arrival, burst out of their eggs and mate. This completes the *Rhabdias* life cycle.

PULMONARY PARADISE

Dissecting the frog's thoracic cavity reveals what appear to be two small blobs of pinkish spit. These are its bubbly-looking lungs. They are disproportionately small because much of the frog's blood is oxygenated through its moist, membranous skin.

I pop a lung with my dissecting needle and tease it apart, carefully avoiding the brownish stains I see inside. A jumble of vibrant life appears—skinny roundworms (*Rhabdias*) writhe and flattish flukes undulate.

The lung flukes, *Haematolechus* and *Pneumonoeces*, feed by sucking in a plug of capillary wall and blood and grinding them up with a muscular pharynx. I can see the remnants of their meal in the gut by using the high power of the microscope. Ground-up pieces of food flow along. But to what end? There is no end. Trematodes have no anus. The anus has not evolved yet. Oh my! The thought is distressing. Gut-inhabiting digenetic trematodes may eat digested food molecules surrounding them. This molecular diet seems not to need an anus. But lung flukes eat bigger stuff. The solution: there is a two-way flow. Wastes are released through the mouth.

I reduce the magnification. I can see the whole worm. Two circles are visible: the oral and ventral (bottom) suckers characteristic of flukes. But the ventral one is partly obscured by a thick, brown, tortuous tube, the egg-filled uterus. Golden-shelled eggs can be seen near the ovary; they become brown as they pass rearward, the brown color signifying that eggshells have been tanned like shoe leather, made impervious to the elements. I can barely make out the transparent empty end of the uterus. It has just expelled dozens of eggs into a space in the lungs.

This whole-animal view reveals that the trematodes have a complex anatomy, almost obscured by the massive uterus that almost fills the body. The worm is a veritable egg-laying machine.

ALL YOU NEED IS A FROG

All you need is a frog. Frogs have everything. As a model to simulate human parasite infections, frogs are the answer. Consider the following symbionts of frogs.

PROTOZOA

Balantidium sp., *Nyctotherus cordiformis* —ciliates in the bowel
Lankestria minima—causes "frog malaria"
Haemogregarina termporariae—gregarine in the blood
Isospora lieberkuhni—sporozoa in gut
Trypanosoma pipientis, rana, etc.—(9 kinds of) trypanosome
Trichomonas spp.—flagellates in the gut
Opalina

NEMATODES

Foleyella (2 species)—microfilaria larvae in blood
Rhabdias bufonis, rana—in lungs
Capillaria xenopodis—intestinal nematode

MONOGENETIC TREMATODE

Polystoma integerrimum

DIGENETIC TREMATODES

Haematoloechus medioplexus—lung fluke
Haplometra cylindracea—lung fluke
Pneumonoeces variegatus—lung fluke
Glypthelmis hyloreus—intestinal fluke
Clinosotomum marginutum—metacercaria
Diplodiscus subclavatus —metacercaria
Ribieroa ondatra—metacercaria

CESTODE

Spirometra erinacei—plerocercoid
Hydatid cysts (unknown species)

ACANTHOCEPHALAN

Pseudoacanthocephalus betsileo

MAGGOT (FLY LARVA)

Batrachomyia sp.

(This is just a brief, superficial list.)

A SAD END

This story ends on a sad note. Frogs began to disappear toward the end of the twentieth century, perhaps as early as the 1980s, when biological supply houses began having difficulty supplying them. In order to perform their noble role as models of living systems, it was necessary to collect them from their millions in massive marshes. Lab raised animals would not do. To be useful, frogs have to be exposed

to the realities of the parasitic world, the worms and protozoa perme-
ating the mud, the fly and mosquito vectors waiting to inject larval
parasites.

Mid-twentieth-century field researchers in tropical rain forests, the
moist, warm havens for hundreds of amphibian species, began to report startling declines in population sizes. Soon species disappeared
in local mass extinctions. Then the disaster spread to more-northern
swamps, and our frogs reached a level of local extinction. The supply houses were unable to provide the northern leopard frog, *Rana
pipiens*, so they found populations of southern leopard frogs, *Rana
sphenocephala*. Then these too almost disappeared. You can still buy
live frogs, but at an exorbitant price. The ubiquitous frog of the fresh-
man biology class is no more.

What has caused this frightening precipitous decline in species diversity and populations? Perhaps the same factors that have caused
a similar worldwide decline and extinctions of coral reefs. Frog and
coral disasters are worldwide. What universal factor might be affecting
populations in this manner?*

Sad.

---

*I imagine you have an idea: climate change. There have been many hypotheses
for this catastrophic frog extinction, from increased UV radiation to habitat change to
foreign fungus associated with increased rainfall or desertification, but all are related
to climate change.

PLATE 29

A. NORTHERN LEOPARD FROG, *Rana pipiens.* To 4 1/3 inches; green; brown spots on back, sides, legs; underside white, two white lines on back; male smaller than female, has thick thumb to grasp female when mating; no internal fertilization; male sprays sperms on eggs as female lays them. Host to at least fifty species of symbionts, including:

B. FROG LUNG FLUKE, *Haematoloechus* spp. To 3/8 inch; in frog lung with other lung fluke, *Pneumoneces medioplexus.* Eats blood; tiny ventral sucker; oval below it is ovary; two ovals beneath, testes. Cercaria swims to dragonfly larva, penetrates, becomes encysted metacercaria. Larva undergoes metamorphosis, retaining cysts. When adult dragonfly is eaten by frog, meta cercaria enters lungs. Eggs laid in lungs carried to throat, where swallowed; pass out in feces, eaten by snail, eventually producing cercariae.

C. OPALINA, *Opalina ranarum.* Ciliated protozoan visible to naked eye; only in rectum of frogs and toads; flattened, many nuclei, no mouth; absorbs molecules from rectal fluid; commensal; not a ciliate, related to amebas.

D. FROG BLADDER MONOGENETIC TREMATODE, *Polystoma integerrimum.* To 1/4 inch; four eyes; opisthaptor has six round suckers, two anchors, sixteen hooks. In bladder of frogs, toads. Only "internal" monogenetic trematode.

E. FROG ACANTHOCEPHALAN, *Pseudoacanthocephalus betsileo.* White; spiny proboscis embedded in upper intestinal wall.

F. FROG LUNG ROUNDWORM, *Rhabdias bufonis.* To 1/4 inch; eats frog, toad blood; males produce sperms then become females, saving sperms for fertilization; eggs hatch into free-living larvae that penetrate skin of frog. Male worm shown.

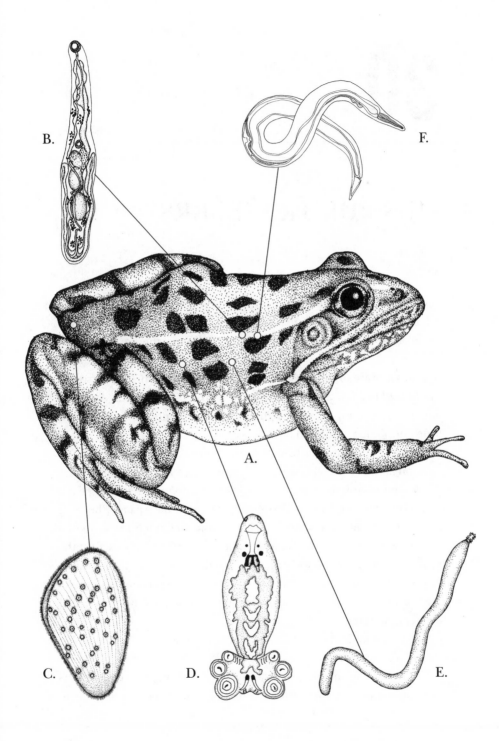

B.

F.

A.

C.

D.

E.

30

TIPS FOR TRAVELERS

I HAVE MENTIONED that many physicians (except specialists in tropical medicine) are not trained to diagnose parasitological diseases. In most cases they treat the symptoms and allow the parasitemia to take its course. In this era of international travel, the potential patient should be aware of the possibility of an exotic infection. Of course this would lead to a flood of recent returnees from Africa, South America, and East Asia bringing self-diagnoses to harried internists, a case of a little knowledge going too far.

I offer these examples from my own experience:

Soon after returning from a rain forest, I developed a progressively reddening pimple near my nose. Close examination revealed that a tiny tube extended from it. A friend had visited the area and become afflicted with cutaneous myiasis. Her pimple contained the larva of a bot fly, *Dermatobia hominis*. A dim recollection led to a little thrill at the thought of having yet another parasite to describe to my students first hand. I called my dermatologist and suggested he prepare his camera for the first case of (what I assumed was) cutaneous myiasis he had ever encountered.

He took one look and told me it was a common pimple and that he had removed dozens. He cut it off. "Why does it have a tube with a distinct hole in the end?" I insisted. He just shrugged his shoulders and looked wise and weary. On examining the specimen that was to be sent to the pathologist for routine examination, I didn't see a maggot. None was described in the subsequent report. I still insist I am right. Why would a pimple have a tube?

Then there is the case of my medical-school fiasco. The parasitology class I took consisted of a never-ending procession of prepared, stained slides with lectures on symptoms. One day we studied *Giardia lamblia*, an intestinal parasite that causes diarrhea. My baby daughters were constantly spreading diarrhea pathogens to us. I decided to check out my kids' feces. Peering through the scope, I saw two eyes staring back at me! *Giardia* has two nuclei with big black spots in them (karyosomes), causing them to resemble eyes. I was horrified and at the same time proud that I had diagnosed my first fecal parasite. A little pompously, I showed my slide to an expert at the medical school.

What I had seen was an abundance of the harmless human intestinal yeast cells—coupled with a vivid imagination.

A case of a little too much knowledge.

EYEBALL TO EYEBALL

If you return from a trip abroad to find you have projectile vomiting, roaring flatulence, sulfurous belching and explosive diarrhea, the bad news is that you won't die; you just have an attack of giardiasis, a form of purgatory devised by the single-celled parasite known as giardia [sic].

—NICHOLAS WADE

Larry was an unusual student. He was particularly creative. Instead of backpacking through Europe like the rest of the middle-class kids in his graduating class, he went off to Africa. He was hiking in Masai country in Kenya when his buddy challenged him to climb Mount Kenya. With youthful bravado he accepted. It was hot at the base of the mountain and cold near the top. He had not thought of preparing for the hike. He wanted to live off the land, pluck fruit off the trees, and drink from the streams. He found no food but did drink from a stream.

Upon his return to the United States, he discovered a surprise souvenir—diarrhea. This condition persisted for *five years*, despite many visits to physicians. Forgetting my previous protozoan embarrassment, I diagnosed the condition as giardiasis. *Giardia* swims through the intestinal river by means of eight undulating, hairlike flagella, adhering to the gut wall with two suckerlike disks. Between the disks and the flagella, it is impressive for a one-celled organism. Behind the disks are two eyelike nuclei that, combined with its oval shape, make it resemble a cute little elf's face. The "elves" adhere to the intestinal walls in huge numbers. This interferes with the absorption of food molecules and irritates the gut, increasing its motility. The intestinal contents are flushed through the gut so fast that water is not absorbed adequately to produce formed stools. Diarrhea results.

The parasites enter the body as cysts in contaminated water. Many species of mammals can release millions of cysts in each defecation. Perhaps Larry was anointed by cysts from a hyena!

Giardia divide longitudinally in their cysts. Once in the intestine, two juveniles escape from each cyst. Free in the all-providing gut, they divide wildly. It has been estimated that a single diarrheic stool can contain fourteen billion cysts, whereas a stool in a moderate infection may contain three hundred million."[35]

As the fecal fluid begins to dehydrate and solidify in the colon, encystment occurs. Vast numbers of these time-resistant cysts can contaminate towels and home-made lemonade, and it is likely that the family will become infected. Children often harbor *Giardia*, the most common human intestinal flagellate. It is so common, for example, that I have never met a person who visited Moscow who did not return with giardiasis. Of course, I don't know that many tourists from Moscow.

Larry eventually recovered.

MEDICAL MACARONI

The canoe tipped slightly as my colleague aimed his pistol at the black water snake *Natrix sipedon* undulating across the surface of the tan, opaque water. Bang! The snake instantly went limp. But there was no wound. The cartridge contained dust shot, minute particles of lead designed to kill and leave no trace, like a neutron bomb.

I ran my hand down the scaly skin of its handsome body and felt distinct longitudinal bulges. Dissection revealed several pieces of what seemed to be macaroni under the skin. I dissected an inch-long specimen, and it proved to be as flaccid and featureless as real macaroni. Later I discovered a minute scolex buried at one end, invisible to the uninitiated. This was the plerocercoid of an aquatic tapeworm previously discussed. It has another name, sparganum, and the human disease it causes is sparganosis. It was common in China in the nineteenth century and was thought to be some sort of new parasite. Someone created a name for it, *Sparganum mansoni*. It was not until early in the twentieth century that it was discovered to be a developmental phase of the life cycle of aquatic tapeworms like the gefilte fish tapeworm, but the name stuck.

The first larval stage of this terrible tapeworm and its relatives is a microscopic ciliated sphere taken in by a copepod, where it undergoes further larval development. If a human drinks contaminated water and ingests the copepod, the larva escapes in the intestine as it should, "thinking" that it is in its next host, a minnow. Following its genetic directive, it migrates to the muscles and encysts, waiting to be eaten by a walleyed pike or other appropriate second intermediate host. To the worm the human is just another fish. But it is destined to live out its larval days buried in muscles or organs of the human— unless the unfortunate person is eaten by an unfortunate bear, the definitive host. But the sparganum does not confine itself to life under the skin. It can migrate throughout the body, encysting anywhere— the liver, the brain, or the eye. That rare occurrence is what makes it so dangerous. To this day it is relatively common in China and Korea, where ocular sparganosis is found. A native remedy for eye infections is to place a dead snake on the affected eye—the cure is the cause! Dead snake is also the cure for an inflamed vagina.

Folks in these countries also eat raw snake for some reason. (Millions of water snakes are harvested in one region of China, where a huge lake is overfished and snakes are an alternative source of protein.)

In the United States, the current cancer craze causes an awareness of subdermal lumps, and more than one physician has been shocked to find a gleaming, white worm in a lanced nodule.

BOTS

I removed a recently dead meadow vole, *Peromyscus gossipinus*, from the trap and picked him up by the tail. He twisted in the wind and his underside swung toward me. A small hole was visible near his crotch. Upon dissection I found an inch-long, caterpillar-like, fat gray spiny maggot of *Cuterebra emasculator.* This ugly, huge fly larva is always found near the crotch, and it often lives up to its species name. A human is more than twenty times the length of a vole or mouse. That would make the maggot—there must be something wrong with my calculations—the equivalent of a twenty-inch-long maggot near the human crotch! The makings of a horror movie!

A member of the same family, the human parasite, the bot fly, *Dermatobia hominis*, is about twice the size of a housefly and produces a more modest larva.* Oddly, it relies on mosquitoes, biting flies, and even a species of tick to disseminate its eggs. The fly is one of those "buzzers" and makes a loud noise, making it vulnerable before it can lay its eggs. Most parasitic flies are sneaky. (The tsetse fly silently lands on you, then bites and flies off before you know it was there.) *Dermatobia* solves the problem by catching a mosquito and laying eggs its abdomen. The mosquito takes a blood meal. The fly's eggs immediately hatch. The maggots fall onto the host's skin, where they promptly drill themselves into the bite wound and end up in the blood-rich dermis. They dig out a little capsulelike haven under the skin, leaving a tiny hole from which they extend their posterior "nostrils," two spiracles. It is this tiny lesion that leads the tropical traveler to her dermatologist. "I am too old for a zit," she tells him. He peers at it and sees the characteristic bulge of the maggot. Puzzled, he makes an incision, sees the larva, and removes it. Case closed.

Here is an ad hoc method of removal: "A bot fly can be recognized in early stages by a small pinhole in your 'mosquito bite.' The easiest

---

*The closest definition of bot is "one who sponges off another." I doubt that this is the origin of the use in botfly. Another definition, from the Dutch, is a derivative of the word "butt." Perhaps the originator was bitten on the butt by this fly.

way to remove a bot fly larva without medical attention is to fill a soft drink bottle with smoke and then hold it tightly to the opening. The larva will begin to suffocate and partly exit . . . allowing the host to squeeze it the rest of the way out. Do NOT PULL . . . as it is anchored in with a series of hooks."[36]

The usual, if less imaginative, method is to apply Vaseline or nail polish to the opening. It will suffocate the maggot and make it back out.

ENGLISH CUISINE

I don't mind eating haggis, bangers and mash, and even kidney pie when in Britain, but I draw the line at watercress sandwiches. Lucky for me.

The giant of the fluke world lives in the bile passages of the liver. The inch-and-half-long, leaflike body of *Fasciola hepatica* is sustained by eating liver cells and blood, causing an inflammation that can lead to excessive scarring (cirrhosis).

It infects its definitive hosts, cattle and sheep, when they graze on underwater vegetation while drinking at a stream. The first larval stage of the worm leaves the snail intermediate host, swims around, and encysts on water plants, becoming the infective metacercaria. There it waits in its cyst for a victim.

Deer also eat underwater vegetation. Every once in a while, deer hunters are asked to remove the liver at a conservation department weighing station. The game warden slices the liver into inch-thick slices for examination. In a recent survey in New York, 17 percent of the deer were infected.

No matter if you like liver or not, fascioliasis is one of the most important diseases of domestic sheep, goats, and cattle in the world and a tremendous economic burden.

Unwashed or poorly washed watercress and other "tempting" aquatic plants can transfer the encapsulated metacercarias to humans, where they excyst in the gut and end up in the liver, causing nasty symptoms like necrosis (tissue disintegration), and possibly poisoning the host with their metabolites. A species of *Fasciola* has been known

PLATE 30

A. This woman in Rembrandt's *The Woman with the Arrow* (1661) has dropped the arrow and appears to be weakly waving goodbye. A fitting ending for this book. Infecting her colon is the protozoan, *Giardia lamblia*; her liver is damaged by the fluke, *Fasciola hepatica*; lesions in her skin contain the flesh-eating larva (maggot) of the bot fly *Dermatobia hominis*. No wonder she is waving so weakly.

B. *Giardia lamblia* microscopic protozoan; eight flagellae, two nuclei with black dots (karyosomes); suction-cup bottom. Millions coat walls of colon; impair absorption, cause diarrhea.

C. GIANT LIVER FLUKE, *Fasciola hepatica*. Largest human trematode parasite; to 1¼ inches long by 5/8 inch wide; brown; in bile ducts and liver. Can cause cirrhosis of liver. Dots anound edge, yolk glands; "tree" testes; convoluted tube, uterus with ovary at base.

D. MAGGOT (larva) OF BOT FLY (TORSALO), *Dermatobia hominis*. To 1 inch; 3/8 inch wide; white with black spines; two elliptical openings at narrow end project from hole in skin act as nostrils; mouth is dot at bottom tip of swollen body. Causes large bump or mosquito bite-like lesion with hole in tip. Mature maggot visibly moves under skin. Drops to ground to pupate after two to eight weeks in host.

E. EGGS OF BOT FLY. Fly captures mosquito in midair and deposits twelve to thirty eggs on abdomen (encircled on plate). When mosquito bites, eggs hatch and tiny maggots crawl to bite site and burrow into lesion.

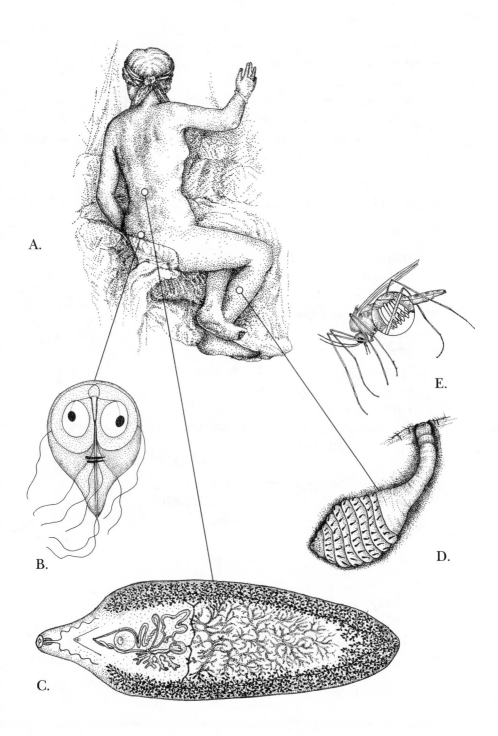

A.

B.

C.

D.

E.

to kill elephants. In a survey of the sheep-farming area of Bolivia, 38 percent of the children were infected; the disease is rare in the United States and Britain.

If, by accident, you have eaten poorly cooked liver and see a fluke-shaped space in the just-eaten piece, have no fear. The parasite is only infective as an encysted metacercaria on an aquatic plant. It has become a bit more protein in your diet.

ELEVEN PRECAUTIONS TO TAKE

Here are some precautions to take on your next trip.

1. Don't eat raw or undercooked pork, beef, fish, crabs, pigeons, or snakes.
2. Don't drink water with copepods in it.
3. Don't eat warmed-over noodles in Thailand.
4. Avoid being bitten by mosquitoes, sandflies, botflies, and big bugs.
5. Do not swim in the Nile. If you insist on swimming, stay away from children peeing in the water.*
6. Keep raw, dead snakes away from eyes or vagina.
7. Stay away from ladies of the night. If you insist, keep a can of kerosene by the bedside.
8. If you appear pregnant and haven't had sex in a really long time, check with an abdominal surgeon.
9. Do not let your grandmother taste raw gefilte fish.
10. Avoid eating mouse and rat feces. Avoid leeches and vampire bats.
11. Above all, don't eat watercress sandwiches.

Oh yes, and find a tropical medicine clinic in the nearest big city if you suspect a parasite infection.

---

*This recommendation suggests that there might be snails shedding cercariae into the water in the vicinity of the children. In fact, the cercariae can be anywhere. It is unwise to swim in slowly moving fresh water anywhere in Africa.

EPILOGUE

WHY DO LITTLE OLD JEWISH WOMEN SUFFER FROM HUGE TAPEWORMS? Why do little Egyptian boys produce orange urine? What bad things can come from a kiss (from a big black bug) and a bite (from a louse)? If you have been amused and affected by this book, it will have achieved its putative purpose. But there are deeper issues. Hopefully you were drawn beyond your superficial curiosity.

The science in this book may have seemed daunting. After all, you were challenged by the intricacies of host-parasite relationships, the ecology and evolution of parasites, and immunological dilemmas. But if you have gone beyond the crotch crickets and vulva vermin, you realize that I have tried to draw you in by presenting the material as a human experience—to go beyond the factual and almost inadvertently internalize ideas. If you share with me the awful feeling I had when I realized that the huge worm in the toilet came out of my intestine, or my horror when the pediatrician showed me the pinworm in my daughter's anus, you will want to follow the worms' passage through the body.

Did you feel a slimy sensation of a leech hanging from your neck or an itchy feeling when you read about lice? These images were an attempt to anchor your involvement in reality. In my view, reality is the foundation of learning. Few nonprofessional readers are drawn to abstractions. Most are grounded in the nitty-gritty of life.

You selected this book because the topic interests you. The presentation, in the form of fun stories, titillated your curiosity and you read on, hopefully ready to deal with complexities of life histories that seemed uninteresting at first. The beautiful plates added a further note of reality.

TEACHING REALITY

This is an extension of what I have learned trying to motivate university classes during almost a half-century of teaching. If the student is fascinated, he or she will accept huge, seemingly impossible tasks. In this acceptance is planted the seeds of success in all endeavors. So I suggest that subject matter is but a vehicle for teaching self-confidence. This term, "self-confidence," in the sense used here means willingness to accept what seems to be an overwhelming challenge (like life) and have the feeling that it is possible to deal with it.

This thesis is manifested in my challenge to the student to accept what in essence is two courses in parasitology in one. I am able to get away with this unreasonable demand because I promise to supply the most fundamental and powerful motivation—*fascination*.

When given the choice of selecting a course, students are torn between an urge to simplify a difficult existence as a biology major and take an "easy" course—or to take parasitology. Those with an inherent curiosity cast aside their intellectual fears and choose the hard way. Thus I am assured of a class of self-selected scholars willing to endure responsibilities they never have faced before. But the bargain has two facets. The student agrees to do what seems to be overwhelming—but I must provide the fascination. I do this with *reality*. During the student's fifteen previous years in school, this has been in short supply. Sitting in class hearing about discoveries made by someone else gets stale when, year after year, instructors drone on about abstractions: DNA, mitochondria, viruses, polyunsaturated fatty acids, the citric acid cycle . . . ad nauseum. What is a DNA molecule?* Can you touch it? Can you feel it? Can you watch it do its thing?

But a stinking, threatening, fast-moving cockroach is real. It assails the senses. It evokes dread and loathing. That is reality. Struggle with that, is the message. You don't like it—deal with it. This is where the fascination comes in. The student is starved for reality. Not everyday reality, but intellectual reality. "What will we find inside the gut of the roach?" is reality. Dissecting the gut of a long-preserved cat or fetal pig is not reality. The pig's stink is sterile, pungent, irritating, artificial—formaldehyde or alcohol. The rubbery texture of muscles, the flaccid artificiality of empty intestines are as interesting as the

---

*This is not to say that teaching molecular biology is not important. I suggest that the curriculum should be leavened with experimental organismal biology.

overboiled chicken in chicken soup. No, by this time the student is sick of playing with models; the dead pig is merely a poor model of a living system.

Given the artificiality of his or her previous learning experiences, the student craves not more theory, but the opportunity to apply the theory. This is valid in any intellectual area, from learning language to finagling with finance.

It is exhausting to provide students with reality. But one cannot really teach the marvelous interaction between host and tapeworm without feeding tapeworm larvae to rats. Can you envision the insufficiency of a dry classroom description in contrast to teasing apart the lung of a frog to discover writhing lung flukes inside? Is seeing the excitement engendered when a student sees a huge ciliated protozoan burst out of the roach gut and sail across the field of light under the microscope worth the effort? It takes monumental effort to transport malaria-infected mice luxuriously by car to the lab. Is seeing the student's frenzied research to save the sick mouse worth the effort? Is seeing a student stretch her sense of dedication to reach into a repulsive colony of cockroaches to remove one for her personal research worth the effort? Is watching the excitement and pride in the face of the student when he, of all people in the class, was perceptive enough to discover a mated pair of schistosomes entwined in perpetual embrace worth it?

Fascination. That is my part of the bargain. Their part is to write several detailed scientific papers *and* pass traditional content midterm and final exams. "Impossible," they say. The responsibility seems overwhelming. But in forty years of giving such a course, all but a handful of students have fulfilled all their responsibilities.

THE EVOLUTION OF A TEACHING PHILOSOPHY

I call this a "live" course, live in the sense that it is based on the study of living things (the definition of biology), and live in the sense of being lively.

I am embarrassed to admit I am still tied to abstractions. My lectures are traditional. The class is bombarded with details. I can't bear not to describe the life cycles of what seems to be an infinite variety of symbiotic relationships. I admit it is boring, but I have not evolved sufficiently to create a curriculum that frees my need to "cover" the infinite details of the discipline. Lecture is the appetizer, but the living main course is waiting in the laboratory.

I am a transitional form, a step in the right direction. I know what to do, but I can't do it—one of the many evolutionary experiments toward the next step: the *total reality course.* But, in climbing out of the muck of a sea of facts, I offer the students the (mandatory) opportunity in lab to do experiments, to see living parasites in situ; to conjecture about why they are found where they are found; why some animals have more kinds of parasites than others; why it is virtually impossible to keep the mouse malaria victim alive. This is the second course that I add to the traditional lecture.

Of course, reading even exciting stories is no substitute for studying living parasites. When you read the previous chapters you were supposed to be fascinated and want to learn more, to go beyond the introductory story. It will be a measure of my success (and perhaps yours) when you look at the glossary and find you are familiar with many of the words and concepts.

The ultimate learning experience is to physically feel the subject. To touch, to smell, to hear, to see it. That is why I established a field station in the Caribbean, so my students could literally be immersed in the subject. But I can't bring you to the vast impoverished world that is dominated by the dark shadow of parasitism. This book has been an attempt to help you feel the depths of the subject.

GLOSSARY

Ameba	protozoan characterized by amorphous shape
amebic dysentery	parasitic intestinal disease caused by *Entamoeba histolytica*, characterized by copious diarrhea
antibody	produced by body's immune system in response to a foreign protein (antigen), often binds to the surface of pathogens, killing them
anticoagulant	prevents blood clotting. See hirudin
anthelmintic	pharmaceutical that kills infective worms (nematodes, trematodes, cestodes)
arachnid	arthropods with eight legs; body in two sections; ticks, mites, spiders
bilharzia, bilharziasis	original term for schistosomiasis; sometimes refers specifically to urinary disease caused by *Schistosoma haematobium*
black smoker	conical opening in deep sea bottom emitting superheated mineral-rich seawater
brood parasite	bird that lays its eggs in nest of another species; the young steal sustenance from host bird's offspring
broad tapeworm	*see* gefilte fish tapeworm
buboe	inflamed, swollen lymph glands; purplish swellings under arms indicate bubonic (black) plague; project from back of neck

in cases of fulminating sleeping sickness caused by *Trypanosoma b. rhodesiense*

cecum	blind sac at base of human large intestine from which appendix protrudes; any blind sac protruding from intestines, stomach
cercae	vibration sensing ("hearing") organ at posterior of some insects
cercaria	tailed, free-living infective larval stage of digenetic trematodes
Chagas' disease	caused by *Trypanosoma cruzi*; affects nervous system, heart; often fatal
chlorophyll	green pigment in plants and protozoa that carries on photosynthesis
cilia	microscopic hairs used for propulsion by ciliate protozoa, and *Opalina;* also used to move minute things like dust, contaminants, from human throat
cirrus	intromittent organ in flukes and tapeworms that everts to transfer sperms
cleaning shrimp	mutualistic shrimp that remove parasites from fishes. Small fishes can also be cleaners.
commensalism	relationship between two species where one benefits and the other is not harmed; if host is used only to carry commensal, process is called phoresis
copepod	most common tiny aquatic crustacean; free-living or parasitic; used as food by microscopic predators to whales; frequent vectors

copulatory spicule	paired needle-shaped projections near tail of male nematodes; forms a bridge that allows ameboid sperms to enter vagina
cutaneous leishmaniasis	disease caused by *Leishmania tropica* producing skin sores
cuticle	flexible waterproof covering of nematodes
cyst	protective sphere used by protozoa to survive outside host; chamber in host's tissues containing dormant juvenile or larva
cysticercoid	bladderlike tapeworm larva with invaginated scolex and a "tail," often found in insects
cysticercus	bladderlike tapeworm larva with invaginated scolex; in beef and pork tapeworms; causes the disease cysticercosis
definitive host	host in which parasite reaches sexual maturity
diapedesis	passage of white blood cells through walls of capillaries to site of inflammation
digenetic trematode	parasitic flatworm (fluke) with (usually) two suckers, no anus; infective stage is cercaria
dinoflagellate	aquatic protozoan; free-living stage has two flagellae; mutualistic forms with chlorophyll found in corals are called zooxanthellae
ecology	scientific study of the relationship between organisms and their environment;

	the environment of parasites is the host's body
ectoparasite	external parasite, e.g., lice, fleas, flies, monogenetic trematodes
elephantiasis	gross enlargement of scrotum or extremities caused by filarial worms, notably *Wuchereria bancrofti*
enzyme	molecule that facilitates chemical reactions, e.g., digestion
epithelial layer	tissue lining inside of internal organs; intestinal mucosa is made up of epithelial cells; may have secretory function; outside of skin of fishes
espundia	local South American name for mucocutaneous leishmaniasis; caused by *Leishmania braziliensis*
evolution	process of change in species over time of heritable traits
excyst	to escape from cyst, often occurs in small intestine
filarial worm	causes filariasis, usually by release of microfilaria larvae in blood or blockage of lymph spaces by adults
first intermediate host	usually first of several hosts (usually two) required for larval development in life cycle of nematodes, trematodes, tapeworms
fixed macrophages	stationary white blood cells; usually line passages of lymph nodes, spleen; remove pathogens
flagellated protozoan	moves by means of one (trypanosomes) or many (*Giardia*) flagellae

flagellum	whiplike "tail" used for propulsion
gastric cecae	saclike extensions of stomach that increase capacity for storage in leeches and other bloodsuckers
gastric juice	acidic fluid in stomach; contains enzymes
gefilte fish tapeworm	*Diphyllobothrium latum*, huge aquatic tapeworm; plerocercoid larva encysts in yellow pike, people
gonopore	reproductive opening, usually in lieu of vagina or opening of uterus to release eggs
granuloma	lesion characterized by cells containing large granules
gravid	containing eggs
gravid proglottid	proglottid of tapeworm that contains eggs
gynecophoral canal	elongate chamber in male schistosome enveloping female
heartworm	filarial disease of dogs where adult is found in heart; caused by *Dirofilaria immitis*
hemocoel	internal chamber mimicking peritoneal cavity in arthropods
hemoglobin	red pigment in blood cells that carries oxygen
hepatitis	liver infection
hermaphrodite	contains both sexes (monoecious)
hirudin	anticoagulant in leech saliva
hookworm	nematodes *Necator americanus*, *Ancylostoma* spp. inhabit small intestines and

	eat red blood cells, causing hookworm anemia
hydatid cyst	spherical chamber containing larvae of smallest tapeworm, *Echinococcus granulosus*
hydatid sand	free scolexes of *E. granulosus* in hydatid cysts
hyperplasia	enlargement of an organ caused by abnormal increase in number of cells
hypertrophy	enlargement of an organ caused by abnormal enlargement of cells
inflammation	swelling, accumulation of white blood cells, redness due to enlargement of capillaries in response to tissue damage or accumulated toxins
intermediate host	harbors developmental stage before parasite reaches sexual maturity
Jordan rose	skin lesion caused by *Leishmania tropica*
juvenile	first developmental stage in organisms where stage that comes from egg looks similar to adult; in nematodes; also, young of any species
kissing bug	insects of family reduviidae that are vectors of *Trypanosoma cruzi*, e.g., *Triatoma*, *Rhodnius*
larva	developmental stage where infant is dissimilar to adult; first stage after egg in insects
leishmaniasis	disease transmitted by sandfly, *Phlebotomus*, e.g., Jordan rose, kala azar
lumen	space inside a cylinder, e.g., intestinal lumen

lymph	watery fluid. Carries dissolved nutrients, antibodies, dissolved food, gases in separate circulatory system; similar to plasma
maggot	fly larva
malaria	disease caused by *Plasmodium* spp. characterized by periodic high fever and chills
mebendazole	modern anthelmintic; inhibits energy production and intestinal function of worms, killing them
merozoite	stage in malarial cycle; free-living in blood until it enters red blood cell
metacercaria	encysted second-stage larva in digenetic trematodes
metastasize	to spread to other parts of body
miracidium	free-living first stage in trematode life cycle that infects snails
monogenetic trematode	flatworms characterized by posterior fleshy attachment organ (see opisthaptor); primarily ectoparasites of fishes
mucocutaneous leishmaniasis	disease caused by *Leishmania braziliensis* affecting mucous membranes of nasopharyngeal region, called espundia and uta in South America
mucosa	layer of epithelial cells lining inside of internal organs, e.g., intestines
mucus	protein-carbohydrate secretion of epithelial cells called mucous membranes, e.g., in nose, surface of fish skin
mutualism	symbiotic relationship where both organisms benefit

myiasis	infection by fly larvae
necrotic	disintegration of tissues as in "necrotic lesion"
nematode	ubiquitous free-living or parasitic round-worm of phylum nematoda with external resistant cuticle; sexes separate
obligate	symbiotic relationship that is mandatory for survival; in contrast to facultative, in which symbiont normally can survive without needing host
onchocerciasis	parasitic disease caused by *Onchocerca volvulus*; long-term microfilarial infection is called river blindness; vector is blackfly
oncomiracidium	ciliated larva of most monogenetic trematodes
ootheca	egg case in certain insects, mollusks; contains about twenty eggs in cockroaches
opisthaptor	posterior flat attachment organ of monogenetic trematodes armed with hooks, suckers, anchors
parasite	symbiont that benefits itself while harming host in some way; often long-term; virtually always smaller than host
paroxysm	sudden periodic alternating attack of chills and fever in malaria
pathogen	disease-causing organism
peristalsis	rhythmic contraction of intestines, ureters, etc., facilitating transport of contents
peritonitis	infection of peritoneal cavity

pheromone	substance produced in minute quantities that affects the behavior of another organism, usually associated with sexual behavior
photic zone	layer of oceans, lakes where enough light penetrates to facilitate photosynthesis
phytoplankton	usually minute, chlorophyll-containing, aquatic free-floating organisms
pinworm	nematode *Enterobius vermicularis* that causes a disease of the bowel
Plasmodium	protozoan blood parasite causing malaria
pogonophora	a wormy phylum; vent species now considered a subclass, vestimentifera, by some; or a class of annelid worms, siboglinidae, by others; in deep water; marine; lacks digestive system
posterior mesenteric vein	lowest major vein of abdomen, tributaries of which are inhabited by schistosomes
preadaptation	to be prepared to invade a new habitat by evolving in an environment similar to the one to be invaded; sewage-dwelling amebas are preadapted to living in the intestines
predator	animal usually adapted to quickly kill another of a different, usually smaller species
primary host	*see* definitive host
proboscis	any anterior tubular organ used for probing, sucking, food gathering, sensing

proglottid	linear section of tapeworm body arranged by maturity— immature, mature, gravid (not a segment)
protandrous hermaphrodite	contains both sexes sequentially; smaller male becomes a female
proteolytic enzyme	enzyme that facilitates protein digestion
pupa	in insects, developmental stage after larva; usually quiescent
retroinfection	parasitic stage leaves body or intestinal wall, then reenters, renewing infection
river blindness	onchocerciasis. Scars on cornea resulting from long-term infection by filarial worm, *Onchocerca volvulus*
schistosomiasis	disease caused by blood fluke, *Schistosoma* spp.; damages liver, bowel, urinary bladder
scolex	anterior end of tapeworm; armed with hooks, suckers
second intermediate host	required for most tapeworm, digenetic trematode life cycles; harbors larval parasites in cysts (see primary host)
sexually dimorphic	different body appearance of each sex; females usually larger than males in invertebrates and fishes; males often more colorful in fishes and birds
sheath	flexible transparent membrane covering nematode microfilariae
sparganosis	disease where host is infected with tapeworm plerocercoids
species specificity	symbiont is adapted to only one species of host

spermatophore	structure containing sperms; can be placed inside female or deposited anywhere as in some mites
spiracle	opening for passage of air into body in insects; extended from host's skin by maggots
sporocyst	asexual stage of digenetic trematodes in snails; may produce daughter sporocysts, redias, or cercariae; no mouth or gut
stem cell	embryonic undifferentiated cell that can become cells of almost any organ depending on the cellular environment
submucosa	connective tissue and blood vessel layer underneath epithelial lining of gut (mucosa)
symbiosis	living together; any of the three categories: mutualism, commensalism, parasitism, often involving one partner living on or close to another
syzygy	lining up front to back, as in sexual cycle of gregarine parasites
tapeworm	cestode; class of flatworms usually parasitic in small intestines; body divided into head-like scolex; subsequent sections called proglottids
trichinosis	disease where larvae of nematode *Trichinella spiralis* encyst in muscles or other organs
trophozoite	active, feeding stage of protozoa, e.g., parasitic amebas, *Plasmodium*
trypanosome	flagellated protozoan; one flagellum and an undulating membrane; causes sleeping sickness; same family as *Leishmania*

trypomastigote	flagellated stage in life cycle of trypanosomes
typhus	louse-borne rickettsial (bacterial) disease characterized by high fever, vomiting, cough: fatal before antibiotics; aggravated by crowding; different from scrub typhus (tsutsugamushi disease), whose vector is a chigger
vector	any agent that transmits a disease organism; in parasites, often contains a stage needed in developmental sequence
white blood cell	blood cell that fights disease either by engulfing invader or producing antibodies; phagocyte
Winterbottom's sign	swollen lymph nodes on back of neck indicative of fulminating sleeping sickness caused by *Trypanosoma b. rhodesiense*
Zooplankton	animals or protozoa that are unable to swim against wind-borne currents; may be as big as jellyfish or as small as protozoa
Zooxanthellae	mutualistic chlorophyll-bearing dinoflagellate protozoa in corals, jellyfish, sea anemones

SELECTED REFERENCES

1. Kaplan, Eugene H. *Sensuous Seas*. Princeton: Princeton University Press, 2006.
2. Kaplan, Eugene H. *Problem Solving in Biology*. Third edition. New York: Macmillan, 1983.
3. Richardson, Dennis, J. Gauthier, and J. Koritko. "Parasitology Education 2001." *Comp. Parasitol.* 71 (1) (2004): 13–20.
4. Adler, David. "Darwin's Illness." *Israel Jour. Med. Sci.* 26 (3) (1990): 163–64.
5. Roberts, Larry, and John Janovy, Jr. *Foundations of Parasitology*. Sixth edition. New York: McGraw-Hill, 2000.
6. Druilhe, Pierre, et al. "A Malaria Vaccine That Elicits Human Antibodies Able to Kill *Plasmodium falciparum*." *PloS Med.* 2 (11) (2005): e344.
7. McNeil, Donald, Jr. "The Soul of a New Vaccine." *New York Times*, December 11, 2007.
8. Roberts and Janovy. *Foundations of Parasitology*.
10. Crompton, D.W.T. "How Much Human Helminthiasis Is There in the World?" *Jour. Parasit.* 85 (1999): 397–403.
11. McKerrow, J. H., et al. "Anisakiasis: Revenge of the Sushi Parasite." *N. England J. Med.* 319 (1998): 228–29.
12. Deardorff, T. L., S. G. Kayes, and T. Fukumura. "Human Anisakiasis Transmitted by Marine Food Products." *Hawaii Med. J.* 50 (1991): 9–16.
13. Rossi, J. S., et al. "*Anisakis* Larval Type 1 in Fresh Salmon." *Am J. Clin. Pathol.* 78 (1982): 54–57.
14. Noble, E. R., et al. *Parasitology: The Biology of Animal Parasites*. Philadelphia: Lea and Febiger, 1989, p. 238.
15. Cheng, Thomas. *General Parasitology*. Orlando: Academic Press, 1986.
16. Cheng. *General Parasitology*, pp. 438, 439.
17. Kloos, H. K., et al. "Lab and Field Evaluation of a Direct Filtration Technique for Recovery of *Schistosoma* Cercariae." *Am. J. Trop. Med. Hyg.* 31 (1) (1982): 122–27.
18. Kaplan, Eugene H. "*Heterobilharzia americana*, Price, 1929, in the Opossum from Louisiana." *Jour. Parasit.* 50 (1964): 797.
19. Angier, Natalie. "In Parasite Survival, Ploys to Get Help from a Host." *New York Times*, June 26, 2007 (reporting on a study by Janice Moore).
20. Hong, S. J., R. Sawyer, and K. W. Kang. "Prolonged Bleeding from the Bite of the Asian Medicinal Leech, *Hirudo nipponica*." *Comp. Clin. Path.* 9 (3) (1999): 125–31.
21. *Musical Spanish eShop*.

22. Kaplan, Eugene H. "Use of the American Roach in Teaching Principles of Parasitism and Disease." *Jour. Parasit.* 50 (1964): 38.

23. Lehane, B. *The Compleat Flea.* New York: Viking, 1969.

24. Roberts and Janovy. *Foundations of Parasitology,* p. 559.

25. Wilford, John Noble. "Scientists Say Mummies Lice Show Pre-Columbian Origins." *New York Times,* February 7, 2008.

26. McNeil, Donald, Jr. "Jump-start on Slow Trek to Treatment for a Disease." *New York Times,* January 8, 2008.

27. Wall, Richard, J. Stevens, and O. Sattaur. "The Turn of the Screw Worm. . . ." *New Scientist,* June 9, 1990.

28. McNeil, "Jump-start."

29. Wilson, David S. *Evolution for Everyone.* New York: Random House, 2007.

30. Kaplan, Eugene H. *Field Guide to Coral Reefs of the Caribbean and Florida.* Boston: Houghton Mifflin, 1982.

31. Pechenik, Jan. *Biology of the Invertebrates.* New York: McGraw-Hill, 2000.

32. Engman, Joseph. "Pogonophora: The Oldest Living Animals?" *Papers of the Michigan Academy of Science, Arts, and Letters* 53 (1978):105–8.

33. Kaplan, Eugene H. *A Field Guide to Southeastern and Caribbean Seashores.* Boston: Houghton Mifflin, 1988.

34. Kaplan. *Sensuous Seas,* p. 21.

35. Chandler, A. C., and C. P. Read. *Introduction to Parasitology.* Tenth edition. New York: Wiley, 1961.

36. Anon. *Urban Dictionary* (online).

ILLUSTRATION SOURCES

REDRAWN FROM

Beck, J. Walter, and John Davies. *Medical Parasitology*. Third edition. St. Louis: C. B. Mosby, 1981.

Buchsbaum, Ralph, et al. *Animals without Backbones*. Third edition. Chicago: University of Chicago Press, 1987.

Cheng, Thomas. *General Parasitology*. Second edition. New York: Academic Press, 1986.

Markell, Edward, M. Voge, and D. John. *Medical Parasitology*. Seventh edition. Philadelphia: W. B. Saunders, 1992.

Noble, Elmer R., et al. *Parasitology: The Biology of Animal Parasites*. Philadelphia: Lea and Febiger, 1989.

Roberts, Larry, and John Janovy, Jr. *Foundations of Parasitology*. Sixth edition. New York: McGraw-Hill, 2000.

Wallace, R., W. Taylor, and J. R. Litton, Jr. *Beck and Brathwaithe's Invertebrate Zoology Laboratory Manual*. Fourth edition. New York: Macmillan, 1989.

INDEX

Page numbers in **bold** refer to plates.